POLYPEPTIDE - NEW INSIGHT INTO DRUG DISCOVERY AND DEVELOPMENT

Edited by **Usman Sumo Friend Tambunan**

Polypeptide - New Insight into Drug Discovery and Development

http://dx.doi.org/10.5772/intechopen.68638
Edited by Usman Sumo Friend Tambunan

Contributors

Christina Bade-Doeding, Wiebke C. Abels, Alexander A. Celik, Gwendolin Simper, Rainer Blasczyk, Toshio Kamiya, Takashi Masuko, Dasiel Borroto-Escuela, Haruo Okado, Hiroyasu Nakata, Cheng Yang, Chao Zhang, Nevra Alkanli, Arzu Ay, Suleyman Serdar Alkanli, Usman Sumo Friend Tambunan, Mochammad Arfin Fardiansyah Nasution, Ahmad Husein Alkaff

Notice

Statements and opinions expressed in the chapters are these of the individual contributors and not necessarily those of the editors or publisher. No responsibility is accepted for the accuracy of information contained in the published chapters. The publisher assumes no responsibility for any damage or injury to persons or property arising out of the use of any materials, instructions, methods or ideas contained in the book.

First published in London, United Kingdom, 2018 by IntechOpen
IntechOpen is the global imprint of INTECHOPEN LIMITED, registered in England and Wales, registration number: 11086078, The Shard, 25th floor, 32 London Bridge Street
London, SE19SG – United Kingdom
Printed in Croatia

British Library Cataloguing-in-Publication Data
A catalogue record for this book is available from the British Library

Additional hard copies can be obtained from orders@intechopen.com

Polypeptide - New Insight into Drug Discovery and Development, Edited by Usman Sumo Friend Tambunan
p. cm.
Print ISBN 978-1-78923-464-0
Online ISBN 978-1-78923-465-7

We are IntechOpen,
the world's leading publisher of
Open Access books
Built by scientists, for scientists

3,650+
Open access books available

114,000+
International authors and editors

118M+
Downloads

151
Countries delivered to

Our authors are among the

Top 1%
most cited scientists

12.2%
Contributors from top 500 universities

Interested in publishing with us?
Contact book.department@intechopen.com

Numbers displayed above are based on latest data collected.
For more information visit www.intechopen.com

Meet the editor

Usman Sumo Friend Tambunan has been a professor of the Department of Chemistry, Faculty of Mathematics and Natural Science, Universitas Indonesia, since 1987. He has also carried out postdoctoral research at Kansas University, Battelle Memorial Institute, Columbus, Ohio, USA, and in Taiwan to study pharmaceutical biotechnology. Since then, he has been a lecturer for Organic Chemistry and Biochemistry at Universitas Indonesia. He has been appointed for several positions at Universitas Indonesia, such as vice dean I of Faculty of Mathematics and Natural Science (2001–2004) and chief of professor council at the Faculty of Mathematics and Natural Science (2007–2015). Prof. Tambunan's research interests focus on drug and vaccine design using computational methods for avian influenza, Ebola virus, dengue virus, and cervical cancer. His research includes the development of natural products, cyclic peptides, and fragment-based drug design to find better drug candidates to combat these diseases. Moreover, he is also specialized in enzyme stability and bioinorganic chemistry studies of the ligand-enzyme complex. His research has been funded by the Ministry of Research Technology and Higher Education and Universitas Indonesia. For his achievements, he has been rewarded with several prestigious awards, such as Satya Lencana in 2010 and 3rd Best Professor at Universitas Indonesia in 2009.

Contents

Preface

Peptides are the biomolecules that consist of several amino acid chains linked through peptide bonds. This class of biomolecules is commonly found in living matter, and some of which play an imperative role in their regulation and functional activities. On the other hand, compared to small molecules, peptides can be readily synthesized and produced massively in industries at a very reasonable price. As a result, peptides have been extensively applied in every aspect of human life, including cosmetics, electronics, the food industries, and pharmaceutical fields. Ultimately, peptide commercialization has become a major trend, which has grown significantly in recent years.

The development of new drugs to combat a variety of diseases has been leading toward investigations into peptides by scientific researchers due to several advantages over other classes of drugs. The usage of peptides in this field varies widely, such as drugs, vaccines, and even as the basic ingredients of microcapsules. Thus, the development of peptides in medical applications is being extensively studied and could provide a significant impact in solving various public health problems worldwide in the near future.

Viewed from the scientific perspectives described above, the editor believes that the development of peptides in every aspect of human life should be enchanted and accelerated. One way to achieve this is by promoting recent research into peptides and their applications in society, especially by other scientists and researchers who specialize in peptide-based research. As such, this book aims to provide insight and knowledge regarding the current research that has been conducted in this field; hence, this book may encourage the reader to emphasize the importance of peptides in human life and to inspire others to identify the potency of peptides for other applications that still need further investigation. The editor has confidence that every single contribution in this book may stimulate the progression of peptides and their applications in human life.

In this book, some of the compelling applications of peptides in biological and health sciences will be introduced. This book comprises an introductory chapter, along with four chapters presented by several prominent scientists who work in the field of peptide research. In summary, these chapters can be divided as follows:

- The introductory chapter describes the application of peptides in biomedical sciences. It will give brief examples of the peptides that have been used as commercial drugs, as well as give a general outline of peptides and their role as therapeutic agents.
- Chapter 2, written by Abels et al., explains the systematic analyzation of peptide profiles under both healthy and pathogenic conditions, which is imperative to give succession to the immunological approach of personalized therapeutics.

- Chapter 3, written by Zhang and Yang, describes the importance of erythropoietin, an anti-inflammatory peptide, and its derivatives in their roles as tissue-protective agents. Moreover, the benefits and limitations of each peptide when used as therapeutic agents are also explained in this review.
- Chapter 4, written by Kamiya et al., develops a new molecular tool that can selectively express monomers or nonobligate dimers of class A G-protein-coupled receptors (GPCRs), which can be handful and beneficial in the field of immunology.
- Chapter 5, written by Alkanli et al., describes the calcitonin gene-related peptide and its isoform calcitonin-associated polypeptide alpha (CALCA), and their roles in several important diseases such as cardiovascular disease and cancers. Furthermore, the polymorphisms of CALCA gene and its associated diseases are also described in this review.

Prof. Usman Sumo Friend Tambunan
Universitas Indonesia
Faculty of Mathematics and Natural Sciences
Department of Chemistry
Kampus Universitas Indonesia Baru
Depok, Jawa Barat, Indonesia

Chapter 1

Introductory Chapter: Application of Peptides in Biomedical Sciences

Usman Sumo Friend Tambunan,
Mochammad Arfin Fardiansyah Nasution and
Ahmad Husein Alkaff

Additional information is available at the end of the chapter

http://dx.doi.org/10.5772/intechopen.79297

1. Introduction

The application of peptides in the pharmaceuticals and medicinal field is thrivingly emerging nowadays. Over the past few years, the development of peptides such as the major compounds in medicines and biotechnology research has been widely utilized due to its high selectivity, good efficiency, predictable metabolism, and affordable prices, compared to other classes of drugs such as small compounds or natural products [1]. Nevertheless, the extensive bioactivities that peptide class possessed are promising and interestingly worthy to be investigated and developed further in the future.

Peptides are biopolymers which composed of amino acids as the monomers that connected through peptide bonds. The length of amino acids is varying from 2 (commonly known as dipeptide) to 50 amino acids (known as a polypeptide). The composition and order of amino acid sequences determine their properties, both physical and chemical, and also their pharmacological activities [2]. Regarding the structure, peptides can be classified into two categories: linear peptide and cyclic peptide. The linear peptide is prone to be hydrolyzed by exopeptidases and endopeptidases. Thus, lowering its effective half-life in the human body. The peptide cyclization is a familiar technique to increase the peptide stability and effective half-life. However, this treatment is also modifying the peptide bioactivity, either by lowering or raising its activity due to fixed flexibility and conformation, which have to be measured later on by *in vivo* experiments [3].

2. Peptide as therapeutics agents

The development of peptides as the therapeutic agents had arisen since the early 1980s when natural peptides such as insulin and adrenocorticotrophic hormone (ACTH) were isolated widely and became the popular therapeutics approaches at that time. Since then, several synthetic peptides (i.e., vasopressin and oxytocin) and natural peptides (i.e., snake venom) have been introduced and identified for their use in pharmaceuticals and biotechnology fields [4]. As for today, more than 60 peptides have been approved for their use as drugs from Food and Drug Admission (FDA) and widely marketed, with other 140, and 500 peptides are still in clinical trials and preclinical development [1]. Therapeutic peptides have various known pharmacological and biological activities, including an antioxidant, antimicrobial, anti-inflammatory, and antihypertension [5]. Currently, the antiviral activity of peptides is also investigated as well, such as an anti-HIV agent [6].

One of the most significant challenges on the development of peptides as the therapeutic agents is to increase their oral bioavailability. Due to their enormous molecular weight, high polarity, low intestinal permeability, and hydrolysis susceptibility (especially for linear pep-tide), almost all therapeutic peptides are administrated through injection. Although consider-able efforts have been made in the advancement of peptides research, so far there are only eight peptide drugs that have currently sold in the market which orally administrated. The well-known examples of this kind of peptide are linaclotide, an oligopeptide which used to treat irritable bowel syndrome with constipation, and cyclosporine, which widely utilized as an immunosuppressant (**Figure 1**) [7]. To date, two approaches have been established to improve the oral bioavailability of the respective peptides: the first strategy would be the optimization of peptide structure, followed by the lead peptide alteration into its derivate that possessed high oral bioavailability. The second strategy, which involves the development of the orally available peptide backbones, has not obtained the same success as the previous strategy, although the potency of this strategy cannot be underestimated yet [3].

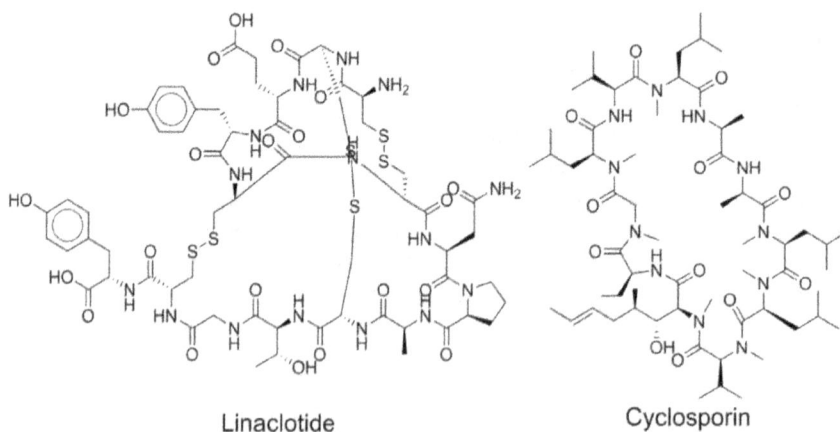

Linaclotide   Cyclosporin

Figure 1. The molecular structure of linaclotide (left) and cyclosporine (right), one of the examples of orally administrated peptide drugs.

Author details

Usman Sumo Friend Tambunan*, Mochammad Arfin Fardiansyah Nasution and
Ahmad Husein Alkaff

*Address all correspondence to: usman@ui.ac.id

Bioinformatics Research Group, Department of Chemistry, Faculty of Mathematics and
Natural Sciences, Universitas Indonesia, Depok, Indonesia

References

[1] Fosgerau K, Hoffmann T. Peptide therapeutics: Current status and future directions. Drug Discovery Today. 2015;**20**(1):122-128. DOI: 10.1016/j.drudis.2014.10.003

[2] Meems LMG, Bernett JG. Innovative therapeutics: Designer natriuretic peptides. JACC: Basic to Translational Science. 2016;**1**(7):557-567. DOI: 10.1016/j.jacbts.2016.10.001

[3] Räder AFB, Reichart F, Weinmüller M, Kessler H. Improving oral bioavailability of cyclic peptides by N-methylation. Bioorganic and Medicinal Chemistry. 2017;**26**:2766-2773. DOI: 10.1016/j.bmc.2017.08.031

[4] Lau JL, Dunn MK. Therapeutic peptides: Historical perspectives, current development trends, and future directions. Bioorganic and Medicinal Chemistry. 2017;**26**(10):2700-2707. DOI: 10.1016/j.bmc.2017.06.052

[5] Sánchez A, Vázquez A. Bioactive peptides: A review. Food Quality and Safety. 2017;1(1):29-46. DOI: 10.1093/fqs/fyx006

[6] Chupradit K, Moonmuang S, Nangola S, Kitidee K, Yasamut U, Mougel M, Tayapiwatana C. Current peptide and protein candidates challenging HIV therapy beyond the vaccine era. Viruses. 2017;**9**(10):1-14. DOI: 10.3390/v9100281

[7] Aguirre TAS, Teijeiro-Osorio D, Rosa M, Coulter IS, Alonso MJ, Brayden DJ. Current status of selected oral peptide technologies in advanced preclinical development and in clinical trials. Advanced Drug Delivery Reviews. 2016;**106**:223-241. DOI: 10.1016/j.addr.2016.02.004

Chapter 2

Peptide Presentation Is the Key to Immunotherapeutical Success

Wiebke C. Abels, Alexander A. Celik,
Gwendolin S. Simper, Rainer Blasczyk and
Christina Bade-Döding

Additional information is available at the end of the chapter

http://dx.doi.org/10.5772/intechopen.76871

Abstract

Positive and negative selection in the thymus relies on T-cell receptor recognition of peptides presented by HLA molecules and determines the repertoire of T cells. Immune competent T-lymphocytes target cells display nonself or pathogenic peptides in complex with their cognate HLA molecule. A peptide passes several selection processes before being presented in the peptide binding groove of an HLA molecule; here the sequence of the HLA molecule's heavy chain determines the mode of peptide recruitment. During inflammatory processes, the presentable peptide repertoire is obviously altered compared to the healthy state, while the peptide loading pathway undergoes modifications as well. The presented peptides dictate the fate of the HLA expressing cell through their (1) sequence, (2) topology, (3) origin (self/nonself). Therefore, the knowledge about peptide competition and presentation in the context of alloreactivity, infection or pathogenic invasion is of enormous significance. Since in adoptive cellular therapies transferred cells should exclusively target peptide-HLA complexes they are primed for, one of the most crucial questions remains at what stage of viral infection viral peptides are presented preferentially over self-peptides. The systematic analyzation of peptide profiles under healthy or pathogenic conditions is the key to immunological success in terms of personalized therapeutics.

Keywords: HLA, peptides, peptide prediction, adoptive T-cell therapies, peptide-vaccination

1. Introduction

The immune system of all species has to be able to discriminate self and foreign (nonself) antigens to combat infections without eliciting autoimmune diseases. The presentation of self and nonself occurs through displaying cellular proteins on the cell surface by proteins of the major histocompatibility complex (MHC) gene cluster. In humans, the MHC locus is termed human leukocyte antigen (HLA) and comprises several gene loci with numerous different alleles for most of the genes [1]. One part of the genes is subsumed as HLA class I (HLA-I) with the gene products being expressed on virtually every nucleated cell in the human body. HLA-I molecules present peptides of intracellular proteins on the cell surface. Cytotoxic T cells (CD8+ T cells) as part of the adaptive immune system can recognize these peptide HLA-I (pHLA-I) complexes by the T-cell receptor (TCR) and scan simultaneously the HLA molecule and the peptide [2, 3] to discriminate between healthy and unhealthy cells, for example, virally infected cells. At the same time, natural killer (NK) cells that are part of the innate immune system scan the cell surface of HLA as well. These cells become activated when HLA-I is missing on the cell surface, for example, on virally infected or tumor cells [4].

Peptide loading on HLA-I is a complex mechanism and determines in addition to the HLA allele which peptides will be presented. The central part of this process is the peptide loading complex that is localized in the endoplasmic reticulum (ER). The HLA-I molecule, consisting of a heavy chain and a microglobulin, is stabilized by several chaperons since the structure is unstable when no peptide is bound. The transporter associated with antigen processing (TAP) imports protein fragments that are degraded in the cytosol by the proteasome into the ER. Depending on the sequence, these peptides are trimmed in different ways in the ER [5, 6]. The bridge between TAP and HLA-I is the chaperon tapasin (TPN) that facilitates peptide binding in the peptide binding groove [7].

Because the HLA gene cluster ranks among the most polymorphic region in the human genome [1] and most of these polymorphisms are located in the peptide binding region (PBR) [8, 9], these polymorphisms result in an abundance of structurally different pHLA entities. In this chapter, we focus on the interplay between HLA alleles, bound peptides and the interaction with immune receptors. It is highlighted that even minor differences in the HLA sequence can impact on the bound ligand or the pHLA structure. Every single peptide changes the overall structure of the HLA molecule. Structural alterations that differ from self-pHLA structures will be recognized by the immune system. Therefore, the last parts of the chapter demonstrate the advantage of established immunopeptidomes for immunotherapies.

2. Peptide selection and presentation

The viability of the immune system is governed by interactions between effector cell receptors and their cognate antigenic ligands. Immune effector cells survey HLA-I molecules on the surface of antigen-presenting cells by indirectly scanning the proteomic content of every single cell. The fundamental role of CD8+ T cells, the elimination of pathogens, is elicited through HLA-I molecules complexed to a peptide of foreign (e.g., viral) origin.

Positive and negative selection of T cells in the thymus is a critical step for the development of a mature functional immune system. Immune cells that have not developed immune tolerance against specific pHLA-I complexes during thymus selection will recognize these antigens as foreign. The allele-specific and patient-specific peptides that are presented on the cell surface shape the individual immune response. Even a single alteration in the peptide sequence can be recognized by immune effectors. Single alterations in the sequence of the HLA heavy chain might not affect which peptide sequences can be bound but could lead to a modified overall pHLA-I structure or might affect the strength of peptide binding resulting in pHLA-I complexes with different half-life times. Besides the influence of amino acid (AA) exchanges within the heavy chain, peptides might undergo competition in patients who carry alleles with the same peptide-binding motif. For those reasons, the presentation of a given peptide is dependent on the HLA type and the health status of the patient. Half-life times of pHLA

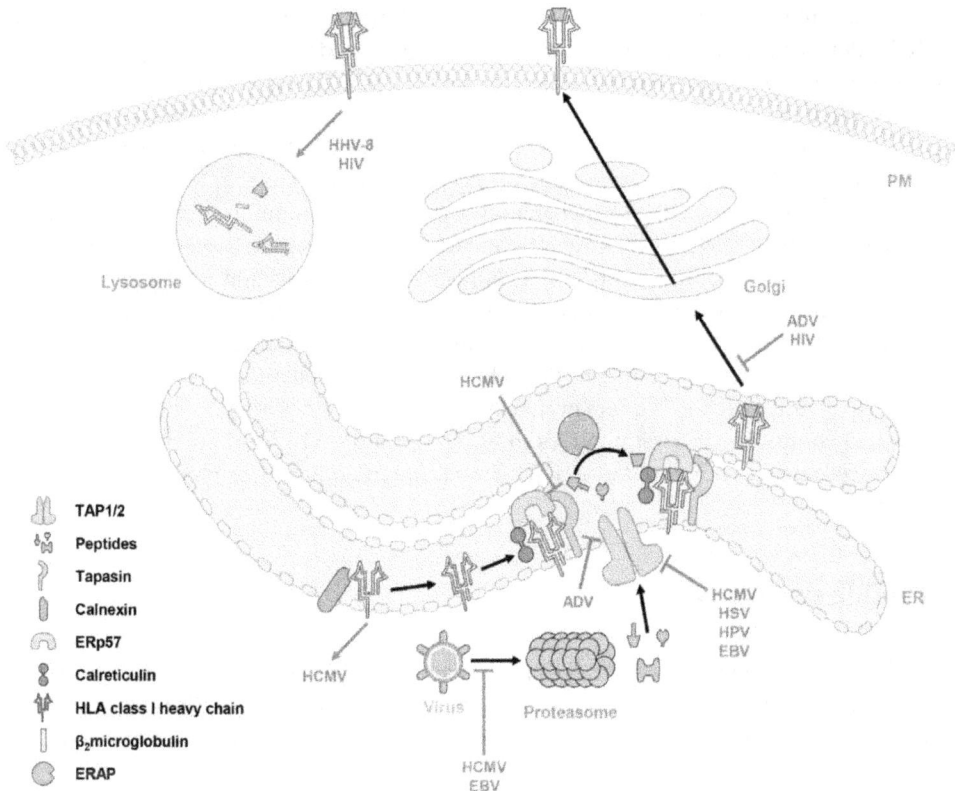

Figure 1. Viral interference with HLA class I peptide presentation. Depicted are targets for viral interference with peptide presentation on HLA-I molecules. HLA-I maturation and surface presentation of peptides are blocked through different mechanisms early postinfection. Most viruses directly target peptide loading through TAP and peptide optimization by tapasin (HCMV [104, 105], HSV [106], HPV [107], ADV [108], EBV [109]). Additionally, cell surface expression is impeded by retention of HLA-I in the ER (ADV [110], HIV [111], HCMV [112]) or rapid degradation of surface molecules (HHV-8 [113], HIV [114]). Other possibilities include dislocation of the HLA-I heavy chain before any peptide loading can occur (HCMV [115, 116]) and inhibition of proteasomal processing of viral proteins by the host cell (EBV, HCMV [117, 118]).

complexes influence the phenotype and/or functionality of T cells. That has to be taken into account when choosing pHLA-I targets for T-cell therapeutics. Based on these facts the importance of knowledge about allelic peptide specificity becomes obvious. Peptides have to pass several intracellular filters and processing steps before being presented to immune effectors cells. The bottleneck is the peptide loading complex (PLC). Not every available peptide in a cell would be necessarily bound to an HLA molecule and displayed at the cell surface since peptides undergo peptide competition during recruitment through the PLC. The PLC consists of several proteins; each has a specialized function for peptide selectivity and specificity. Proteins of this complex are dedicated targets for viral interference and thus viral immune evasion (**Figure 1**).

The HLA heavy chain has to adopt a peptide-receptive form and complex with certain proteins of the PLC. It could be demonstrated that certain allelic variants interact differently with proteins of the PLC and thus are more prone to present peptides of malignant origin. Those HLA subtypes can differ from alleles that are strictly dependent on the association with the PLC for peptide loading only by a single amino acid within the heavy chain, altering the structural interface that interacts with the PLC [10–12]. Especially the interaction of TPN, a protein that mediates the binding of high-affinity peptides into the HLA-I PBR, with the HLA-I molecule, is of exquisite importance to produce stable pHLA-I complexes that persist on the cell surface. However, few allelic HLA variants are able to present peptides without the assistance of TPN. That enables those alleles to continuously present viral peptides at an infectious stage where viral interference with TPN occurs; however, the presentation of self-peptides that did not take part in negative T-cell selection would be facilitated and might lead to uncontrollable autoimmune reactions. HLA-I variants that select and load peptides without the assistance of TPN are likely to present a broad range of low-affinity viral-derived peptides during an infection. However, the presentation of viral peptides during an active viral infection is a rare event since still self-peptides are present intracellularly and would compete with viral peptides to fit into the PBR. The viral peptides that would reach the cell surface complexed to an HLA-I molecule could hardly be predicted by peptide prediction tools.

3. Peptide specificity

Due to the fact that peptides undergo different selection steps before being presented by an individual HLA allele, it has to become clear that the individual HLA profile is the major and most distinguished obstacle. Every HLA allele differs from another by the composition of AAs in and/or outside the peptide-binding groove, resulting in allele-specific profiles of the bound peptides [13–15]. Therefore, the knowledge of an individual peptide binding profile can be used as precedence for the measurement of permissivity between HLA subtypes. The sequence, length and immunogenicity of a given peptide determine the half-life time of the whole pHLA complex and furthermore the specificity and reactivity of their cognate immune receptor.

Several studies demonstrated the impact of the sequence and feature of HLA-bound peptides on receptor recognition. This observation holds true for T-cell receptors of the adaptive immune system as well as for NK-cell receptors of the innate immune system. We recently could highlight the potential of the nonclassical HLA-I molecule HLA-E to select and present peptides of extraordinary length and their effect on differential NK-cell recognition. For the nonpolymorphic HLA-E molecule only two functional variants exist distinguished by a single AA difference. That would imply that HLA-E is, in regard to its peptide profile, invariant. However, by sequencing their bound ligands, we found both alleles presenting a different set of peptides [16]. Since HLA-E is an intermediate molecule for the adaptive and innate immune system, supporting the non-PLC-dependent presentation of peptides during HLA downregulation episodes, the invariability of this molecule would certainly make biological sense. However, the finding that HLA-E subtypes differ in their immunity was somehow unexpected. Reconstitution of empty HLA-E molecules with the designated peptides on the surface of artificial APCs resulted in a peptide-specific immune recognition [17].

Two types of NK-cell receptors interact with HLA-I molecules, killer cell immunoglobulin-like receptors (KIRs) and C-type lectin-like receptors. The former ones are highly polymorphic and polygenic in the population and recognize HLA-A, -B and -C alleles [18], whereas the latter ones bind among others to HLA-E [19, 20]. CD8+ T cells recognize endogenous HLA alleles and assess their immune status by virtue of the presented peptide. The display of different peptides thus allows for precise monitoring of the immune status of the cell through the adaptive immune system. However, this also means that these cells have to be primed on the recognition of specific peptides that are usually derived from endogenously expressed proteins [21]. The presented peptide repertoire can be altered by aberrant protein expression as well as the presence of foreign proteins (e.g., viral proteins) in the cell. To counter recognition by CD8+ T cells, many viruses have developed immune evasion strategies that specifically target HLA peptide loading and presentation. For instance, the HCMV protein US6 interferes with peptide translocation at the ER thus depriving the available peptide pool or even directly procures that the HLA heavy chain is degraded through US2 or US11 [22]. However, in the event of such disrupted HLA-I presentation, NK cells become activated. Although NK cells do not recognize the specific HLA allele, the absence of HLA expression on the cell surface triggers NK-cell activation. Nevertheless, NK cells can still recognize certain peptides in the context of the nonclassical HLA molecule HLA-E that presents a very narrow set of peptides derived from the signal sequence of other HLA class I molecules. These peptides are 9 AA in length and are anchored preferably by Met at peptide position p2 and Leu at pΩ, whereas positions p4, p5 and p6 are accessible to the solvent [23, 24]. On NK cells, HLA-E in combination with these peptides is recognized by inactivating the NKG2A/CD94 heterodimeric receptor complex [25]. In the absence of HLA class I molecules or during cell stress, HLA-E was shown to present noncanonical peptides of different length [16, 17], for example, the Hsp60-derived peptide QMRPVSRVL that causes loss of recognition by NKG2A/CD94 [26] or the HIV Gag-derived peptide AISPRTLNA that causes HLA-E upregulation [27]. In the case of HCMV, a peptide from the UL40 protein that closely resembles the sequence of leader peptides from certain HLA-C allotypes is provided to stabilize HLA-E expression in infected cells. However, in individuals negative for these HLA-C allotypes, the UL40-peptide constitutes the presentation of nonself on HLA-E and can thus elicit a CD8+ T-cell response [19, 28]. Additionally, HLA-E in complex with other pathogen-derived peptides was shown to stimulate CD8+ T-cell responses. For instance, the Epstein–Barr virus-derived peptide SQAPLPCVL was shown to be recognized by the αβTCR of a CD8+-CD94/NKG2C+ T-cell clone [29, 30] described HLA-E-restricted *Salmonella enterica serovar Typhi*-specific CD3+CD8+CD4−CD56− T cells. These diverse interactions demonstrate the subtle interaction of innate and adaptive immunity through the presented peptide on HLA-E.

4. Peptide binding prediction, bioinformatic tools

To identify peptides that would be suitable for application in cellular therapeutic strategies, certain properties have to be analyzed: (1) the peptide-binding motif of the HLA allele of choice and (2) the HLA allele-specific features of the bound peptides such as length and topology. There are several bioinformatic tools that enable scientists and clinicians to predict peptides that would be presented by a certain HLA allele, yet, these tools do not consider allele-specific features and immune dominance of peptides. The kinetics of antigen expression and the competition of peptides to be preferentially bound and presented are also not considered by these bioinformatics prediction tools. Most data available in these tools are based on experimental peptide data (**Tables 1** and **2**). However, it remains unclear if those predicted peptides would ever be naturally presented. Peptides predicted from for example a viral protein would not necessarily be processed, selected and/or presented by the respective patient awaiting T-cell therapy. Therefore, the pathogen- or peptide-specific T cells that would be transplanted might not be able to find their mutual pHLA molecule. An example of the first successful adoptive transfer of virus-specific T cells described the transfer of HCMV-specific T cells and their reconstitution of antiviral immunity in an immune-deficient bone marrow transplant recipient [31]. The technique of adoptive T-cell transfer could be further improved leading to the selection of specific T-cells based on IFN-γ secretion or pHLA multimer staining and selection following antigen stimulation [32–34]. Both techniques bear the imperative to know which viral peptides are presented on the particular HLA subtype of, for example, HCMV-infected cells. So far, few HLA-restricted peptides have been studied. The majority of peptides are derived from the well-characterized phosphoprotein (pp)65 or the immediately early (IE)1 protein, however, not for every patient responses against these two proteins are immunodominant [35, 36]. Best studied are the pp65-derived peptides NLVPMVATV and TPRVTGGGAM, restricted to HLA-A*02:01 and HLA-B*07:02, respectively. Those peptides are described to induce extremely strong T-cell responses [37–42]. Yet, these peptides have been computationally predicted [43] but not been isolated from HLA molecules. Thus, it remains unproven if they would ever be naturally presented. That might be an explanation for the failure of long-term T-cell transfers [33, 44, 45].

Name	Application	Methods	Ref.	Number of HLA class I alleles	Number of HLA class II alleles	Peptide length	Other species	URL
BIMAS	Predicts half-time of dissociation of peptides from protein sequences	Coefficient tables	[58]	41 inc. supertypes	0	8–10	No	https://www-bimas.cit.nih.gov/molbio/hla_bind/
EpiJen	Predicts peptide binding from protein sequence (proteasome cleavage, TAP binding and MHC binding)	Multi-step algorithm	[59]	18	0	9	No	http://www.ddg-pharmfac.net/epijen/EpiJen/EpiJen.htm
hla_a2_smm	Predicts binding affinity of peptides, high affinity HLA-A2 binding peptides from protein sequence and mutated peptides with higher affinity	SMM pair coefficients	[60]	1	0	9–10	No	https://zlab.bu.edu/SMM/
IEDB T Cell Epitope Prediction Tools	Predicts T cell epitopes from proteins (MHC binding, processing and immunogenicity)	Several tools can be chosen	[61–64]	77	n/s	Class I:8–14 Class II: n/s	Chimpanzee, cow, gorilla, macaque, mouse, pig, rat for MHC class I; mouse for MCH class II	http://tools.iedb.org/main/tcell/
Mappp	Predicts antigenic peptides to be processed and presented by MHC class I from peptide or protein sequence	Uses BIMAS or SYFPEITHI for binding prediction	[65]	35 inc. supertypes	0	8–10	Mouse, cattle	http://www.mpiib-berlin.mpg.de/MAPPP/index.html http://www.mpiib-berlin.mpg.de/MAPPP/binding.html
MHC2MIL	Predicts binding affinity of MHC-II peptides from protein sequence	MIL	[66]	0	26	9–25	No	http://datamining-iip.fudan.edu.cn/service/MHC2MIL/index.html
MHC2PRED	Prediction of MHC class II binders	SVM	[67]	0	38 inc. supertypes	9	Mouse	http://crdd.osdd.net/raghava/mhc2pred/index.html

Name	Application	Methods	Ref.	Number of HLA class I alleles	Number of HLA class II alleles	Peptide length	Other species	URL
MHCBN	Database with information about allele specific MHC binding peptides, MHC nonbinding, TAP binding, TAP nonbinding peptides and T-cell epitopes	Database	[68, 69]	n/s	n/s	n/s		http://crdd.osdd.net/raghava/mhcbn/index.html
MHCMIR	Predicts binding affinity and levels of MHC-II peptides from peptide or protein sequence	MIR	[70]	0	13	All	Mouse	http://ailab.ist.psu.edu/mhcmir/predict.html
MHCPRED	Predicts binding affinity of peptides to MHC class I and II molecules and to TAP from protein sequence and calculates binding affinity for heteroclitic peptides	Additive method, partial least square regression	[71–73]	11	3	9	Mouse	http://www.ddg-pharmfac.net/mhcpred/MHCPred/ http://www.ddg-pharmfac.net/mhcpred/MHCPred/pepLib.html
MMBPred	Predicts mutated high affinity and promiscuous MHC class-I binding peptides from protein sequence, epitope enhancement, 1–3 AAs mutation of nonamer peptides	QM	[74]	40 inc. supertypes	0	9	Rhesus macaque, mouse	http://crdd.osdd.net/raghava/mmbpred/
MULTIPRED	Predicts binding of peptides to HLA class I and class II DR supertypes and individual genotypes	Uses NetMHCpan and NetMHCIIpan	[75]	13 supertypes	13 supertypes	8–11 for HLA class I and genotype 9 for HLA class II	No	http://cvc.dfci.harvard.edu/multipred2/index.php

Name	Application	Methods	Ref.	Number of HLA class I alleles	Number of HLA class II alleles	Peptide length	Other species	URL
NetCTL	Predicts CTL epitopes in protein sequences (Cleavage, TAP transport, HLA class I binding)	ANN	[76]	12 supertypes	0	9	No	http://www.cbs.dtu.dk/services/NetCTL/
NetMHC	Predicts binding of peptide to MHC class I molecules from peptide or protein sequence	ANN	[19, 77]	81 (or 12 supertypes)	0	8–14	Chimpanzee, rhesus macaque, mouse, cuttle, pig	http://www.cbs.dtu.dk/services/NetMHC/
NetMHCcons	Predicts binding of peptides to any known MHC class I molecule from peptide or protein sequence	Consensus (NetMHC, NetMHCpan and PickPocket)	[78]	User specified	0	8–15	Chimpanzee, gorilla, rhesus macaque, mouse, cuttle, pig	http://www.cbs.dtu.dk/services/NetMHCcons/
NetMHCII	Predicts binding of peptides to HLA-DR, HLA-DQ, HLA-DP from peptide or protein sequence	ANN	[79, 80]	0	26	variable	Mouse	http://www.cbs.dtu.dk/services/NetMHCII/
NetMHCIIpan	Predicts binding of peptides to HLA-DR, HLA-DQ, HLA-DP from peptide or protein sequence	ANN	[81, 82]	User specified	0		Mouse	http://www.cbs.dtu.dk/services/NetMHCIIpan/
NetMHCpan	Predicts binding of peptides to any known MHC class I molecule from peptide or protein sequence	ANN	[19, 83, 84]	User specified	0	8–14	Chimpanzee, gorilla, rhesus macaque, mouse, cuttle, pig	http://www.cbs.dtu.dk/services/NetMHCpan/
nHLAPred: ANNPred	Predicts MHC Class I binding regions in proteins	ANN	[85]	26 inc. supertypes	0		Mouse	http://crdd.osdd.net/raghava/nhlapred/neural.html
nHLAPred: ComPred	Predicts MHC Class I binding regions in proteins	ANN/QM	[85]	59 inc. supertypes	0		Rhesus macaque, mouse	http://crdd.osdd.net/raghava/nhlapred/comp.html

Name	Application	Methods	Ref.	Number of HLA class I alleles	Number of HLA class II alleles	Peptide length	Other species	URL
PREDPEP	Predicts binding of peptides to HLA class I from peptide or protein sequence	Published coefficient tables	[86]	6	0	8–10 (dependent on the allele)	Mouse	http://margalit.huji.ac.il/Teppred/mhc-bind/index.html
ProPred	Predicts MHC Class II binding regions in an antigen sequence	QM	[87]	51	0		No	http://crdd.osdd.net/raghava/propred/
ProPred I	Predicts MHC Class I binding regions in an antigen sequence	QM	[88]	39 inc. supertypes	0		Mouse, cattle	http://crdd.osdd.net/raghava/propred1/index.html
Rankpep	Predicts binding of peptides to MHC class I and class II molecules from peptide or protein sequence	PSSM	[89–91]	n/s	n/s	Dependent on the allele		http://imed.med.ucm.es/Tools/rankpep.html
svmhc	Predicts binding of peptides to MHC class I molecules from peptide or protein sequence	SVM, uses MHCPEP or SYFPEITHI	[92, 93]	31	0	8–10 (dependent on the allele)	No	http://svmhc.bioinfo.se/svmhc//
SYFPEITHI	Database of MHC ligands and peptide motifs and epitope prediction	Matrix/motif-based, published motifs	[94]	33	6	8–11 for HLA class I; 15 for HLA class II	No	http://www.syfpeithi.de/
TEPITOPEpan	Predicts tissue-specific binding of peptides to MHC class II molecules from peptide or protein sequence	PSSM	[95]	0	50	9–25	No	http://datamining-iip.fudan.edu.cn/service/TEPITOPEpan/index.html

Abbr. HMM = hidden Markov model, SVM = support vector machine, PSSM = position-specific scoring matrix, QM = quantitative matrices, ANN = artificial neuronal networks, SMM = stabilized matrix method, MIL = multiple instance learning, MIR = multiple instance regression, n/s = not specified.

Table 1. Listing of peptide prediction tools available on the web [Accessed November 2017].

Name	Underlying database/data source	URL for matrices/training data
BIMAS	Coefficient tables deduced from the published literature by Dr. Kenneth Parker, Children's Hospital Boston	https://www-bimas.cit.nih.gov/cgi-bin/molbio/hla_coefficient_viewing_page https://www-bimas.cit.nih.gov/molbio/hla_bind/hla_references.html
EpiJen	AntiJen [96, 97], SYFPEITHI [94]	http://www.ddg-pharmfac.net/antijen/AntiJen/antijenhomepage.htm
hla_a2_smm	BIMAS [58], SYFPEITHI [94], data described in Peters, Tong [60]	https://zlab.bu.edu/SMM/
IEDB T Cell Epitope Prediction Tools	IEDB [61], Sette lab, Buus lab, uses diverse predictions methods (see webpage)	http://tools.iedb.org/mhci/download/ http://tools.iedb.org/mhcii/download/
Mappp	BIMAS [58], SYFPEITHI [94], coefficient tables deduced from the literature by Kenneth Parker, Children's Hospital Boston	—
MHC2MIL	Data by Wang, Sidney [98]	—
MHC2PRED	JenPep [19], MHCBN [68]	—
MHCBN	MHCBN [68]	—
MHCMIR	IEDB [61]	—
MHCPRED	JenPep [19]	—
MMBPred	MHCBN [68]	—
MULTIPRED	See NetMHCpan and NetMHCIIpan	—
NetCTL	See NetMHC	—
NetMHC	Trained for 81 HLA alleles including HLA-A, -B, -C and –E, n/s	—
NetMHCcons	IEDB [61]	—
NetMHCII	Data by [19]	http://www.cbs.dtu.dk/suppl/immunology/NetMHCII-2.0.php
NetMHCIIpan	IEDB [61]	http://www.cbs.dtu.dk/suppl/immunology/NetMHCIIpan-3.0/
NetMHCpan	IEDB [61], IMGT/HLA database [1]	—
nHLAPred: ANNPred	MHCBN [68]	—
nHLAPred: ComPred	MHCBN [68], BIMAS [58]	http://crdd.osdd.net/raghava/nhlapred/matrix.html
PREDPEP	Pairwise potential table by Miyazawa and Jernigan [99]	—
ProPred	QMs by Sturniolo, Bono [95]	http://crdd.osdd.net/raghava/propred/page4.html
ProPred I	BIMAS [58] and matrices by Ruppert, Sidney [100] and Sidney, Southwood [101]	http://crdd.osdd.net/raghava/propred1/matrices/matrix.html

Name	Underlying database/data source	URL for matrices/training data
Rankpep	MHCPEP [102], SYFPEITHI [94], GenBank [103]	—
svmhc	MHCPEP [102], SYFPEITHI [94]	http://www.cs.cornell.edu/people/tj/svm_light/
SYFPEITHI	Published literature	—
TEPITOPEpan	n/s	http://datamining-iip.fudan.edu.cn/service/ TEPITOPEpan/TEPITOPEpan.html

Abbr. n/s = not specified.

Table 2. Listing of the underlying databases/data sources for peptide binding prediction.

5. Analysis of naturally presented peptides

The analysis of the individual patient and cell-type-specific immunopeptidome can be realized through sequencing the HLA-bound peptides. It is imperative for all ongoing peptide studies and cellular therapies to find peptides that are (1) naturally presented by the distinct allele, (2) immunogenic for (at best) a public T-cell repertoire and (3) preferentially presented when different peptides are available. A study from Yaciuk et al. showed for example that the peptides isolated from HIV-infected T cells differ from predicted peptides and exhibit different T-cell reactions, factors that have to be considered in designing immunotherapies [46]. That information is only available after immunopeptidome analyses.

In the past, different methods have been applied to answer these questions comprehensively. There are two reliable methods to determine peptide sequences from selected HLA alleles. First, membrane-bound HLA molecules from recombinant single-antigen-presenting cells [47, 48] or from donor cells [49, 50] can be captured by affinity chromatographic methods and the bound peptides isolated and sequenced by mass spectrometry. Second, the most realizable method is the soluble HLA technology [16, 51]. Vectors encoding for soluble forms of HLA molecules (Exon 1–4) are transfected or lentivirally transduced into the cell line of choice. An optional recombinant tag (e.g., V5 tag) engineered at the C-terminus of the protein enables specific purification of the recombinant HLA molecule of choice without the challenge of contamination by cellular-self-HLA molecules. Both methods have been compared by Scull et al. [52] and indicated as an equivalent for the determination of allele-specific peptides. Furthermore, Badrinath et al. [10] could demonstrate that sHLA molecules associate during peptide acquisition with the loading complex as well. These results prove evidence that the use of sHLA technology for understanding allele-specific peptide-binding motifs, the prerequisite for updating peptide prediction databases, is the most time- and cost-efficient implementation.

For the development of tailor-made T-cell-based immunotherapeutic strategies, the identification of tumor-specific HLA ligands is imperative. The production of recombinant sHLA-expressing cells derived from various tissues of malignant origins would guide towards understanding immune dominance through peptide competition. One of the most innovative applications is the peptide fishing from tumor tissue. Immunological tolerance is mediated through T cells that are primed in the thymus by self-peptides. Therefore, the comprehensive knowledge of the HLA immunopeptidome from diseased cells is fundamental for the development of efficient immuno-therapeutic strategies. The presentation of peptides depends on the health state of a patient. During infections, the expression of HLA molecules and thus peptide presentation, including presentation of self-peptides, is diminished through an immune escape mechanism of the invasive pathogen.

6. Peptide vaccination

The treatment of cancer represents a great challenge due to the fact that the vast majority of HLA-restricted peptides differs from tissue to tissue and is dependent on the tumor entity. For that reason, it becomes obvious how fundamentally important the knowledge of the tumor-specific peptidome is. For personalized cancer immunotherapies, the knowledge of naturally presented peptides [53] represents the exclusive possibility for therapeutical suc-cess. The analysis of the mutanome, the proteomic content of a diseased cell, includes the dis-covery of neo-antigens or post-translational-modified peptides and the avoidance of targeting self-antigens from healthy tissue. The results of such individual mutanomes might alter dur-ing the course of tumor progression [16]. In peptide vaccination trials, the use of multiple peptides in combination [54, 55] represents a useful method for targeting all MHC-presenting cells with the peptide of choice. Yet, since the cell type where the peptide(s) bind to cannot be traced, the rates of antitumor immune responses might differ from patient to patient and certain tumor cells where, for example, low MHC expression rates might remain undetected from the immune system. To achieve a comprehensive and precise analysis of presented tumor antigens, the method of antigen discovery and appropriate T-cell assay for knowledge of immunogenicity of the dedicated antigen for vaccination is the key factor [56, 57].

7. Conclusion

Peptide selection and presentation is an exquisite biological and immunological event. Every single peptide is a mirror of the health state of a distinct cell and determines the outcome of immune recognition and responses. For all cellular therapies, the knowledge of the HLA-subtype specific proteome is crucial for the utilization of ligand prediction tools, which have to be implemented where no experimental data are available, yet.

Author details

Wiebke C. Abels, Alexander A. Celik, Gwendolin S. Simper, Rainer Blasczyk and Christina Bade-Döding*

*Address all correspondence to: bade-doeding.christina@mh-hannover.de

Institute for Transfusion Medicine, Hannover Medical School, Hannover, Germany

References

[1] Robinson J et al. The IPD and IMGT/HLA database: Allele variant databases. Nucleic Acids Research. 2015;**43**(Database issue):D423-D431

[2] Germain RN. MHC-dependent antigen processing and peptide presentation: Providing ligands for T lymphocyte activation. Cell. 1994;**76**(2):287-299

[3] Zinkernagel RM, Doherty PC. Immunological surveillance against altered self components by sensitised T lymphocytes in lymphocytic choriomeningitis. Nature. 1974; **251**(5475):547-548

[4] Kiessling R et al. Evidence for a similar or common mechanism for natural killer cell activity and resistance to hemopoietic grafts. European Journal of Immunology. 1977;**7**(9): 655-663

[5] Roelse J et al. Trimming of TAP-translocated peptides in the endoplasmic reticulum and in the cytosol during recycling. The Journal of Experimental Medicine. 1994;**180**(5):1591-1597

[6] Mpakali A et al. Structural basis for antigenic peptide recognition and processing by endoplasmic reticulum (ER) aminopeptidase 2. The Journal of Biological Chemistry. 2015; **290**(43):26021-26032

[7] Zarling AL et al. Tapasin is a facilitator, not an editor, of class I MHC peptide binding. The Journal of Immunology. 2003;**171**(10):5287-5295

[8] Carreno BM et al. The peptide binding specificity of HLA class I molecules is largely allele-specific and non-overlapping. Molecular Immunology. 1992;**29**(9):1131-1140

[9] Bjorkman PJ et al. The foreign antigen binding site and T cell recognition regions of class I histocompatibility antigens. Nature. 1987;**329**(6139):512-518

[10] Badrinath S et al. Position 156 influences the peptide repertoire and tapasin dependency of human leukocyte antigen B*44 allotypes. Haematologica. 2012;**97**(1):98-106

[11] Badrinath S et al. A micropolymorphism altering the residue triad 97/114/156 determines the relative levels of Tapasin independence and distinct peptide profiles for HLA-A(*)24 allotypes. Journal of Immunology Research. 2014;**2014**:298145

[12] Manandhar T et al. Understanding the obstacle of incompatibility at residue 156 within HLA-B*35 subtypes. Immunogenetics. 2016;**68**(4):247-260

[13] Bade-Doeding C et al. Amino acid 95 causes strong alteration of peptide position Pomega in HLA-B*41 variants. Immunogenetics. 2007;**59**(4):253-259

[14] Badrinath S et al. Position 45 influences the peptide binding motif of HLA-B*44:08. Immunogenetics. 2012;**64**(3):245-249

[15] Huyton T et al. Residue 81 confers a restricted C-terminal peptide binding motif in HLA-B*44:09. Immunogenetics. 2012;**64**(9):663-668

[16] Kraemer T et al. HLA-E: Presentation of a broader peptide repertoire impacts the cellular immune response-implications on HSCT outcome. Stem Cells International. 2015;**2015**: 346714

[17] Celik AA et al. The diversity of the HLA-E-restricted peptide repertoire explains the immunological impact of the Arg107Gly mismatch. Immunogenetics. 2016;**68**(1):29-41

[18] Uhrberg M et al. Human diversity in killer cell inhibitory receptor genes. Immunity. 1997; **7**(6):753-763

[19] Hoare HL et al. Structural basis for a major histocompatibility complex class Ib-restricted T cell response. Nature Immunology. 2006;**7**(3):256-264

[20] Brooks AG et al. Specific recognition of HLA-E, but not classical, HLA class I molecules by soluble CD94/NKG2A and NK cells. Journal of Immunology. 1999;**162**(1):305-313

[21] Bjorkman PJ et al. Structure of the human class I histocompatibility antigen, HLA-A2. Nature. 1987;**329**(6139):506-512

[22] Jackson SE, Mason GM, Wills MR. Human cytomegalovirus immunity and immune evasion. Virus Research. 2011;**157**(2):151-160

[23] Braud V, Jones EY, McMichael A. The human major histocompatibility complex class Ib molecule HLA-E binds signal sequence-derived peptides with primary anchor residues at positions 2 and 9. European Journal of Immunology. 1997;**27**(5):1164-1169

[24] Petrie EJ et al. CD94-NKG2A recognition of human leukocyte antigen (HLA)-E bound to an HLA class I leader sequence. The Journal of Experimental Medicine. 2008;**205**(3):725-735

[25] Braud VM et al. HLA-E binds to natural killer cell receptors CD94/NKG2A, B and C. Nature. 1998;**391**(6669):795-799

[26] Michaelsson J et al. A signal peptide derived from hsp60 binds HLA-E and interferes with CD94/NKG2A recognition. The Journal of Experimental Medicine. 2002;**196**(11):1403-1414

[27] Nattermann J et al. HIV-1 infection leads to increased HLA-E expression resulting in impaired function of natural killer cells. Antiviral Therapy. 2005;**10**(1):95-107

[28] Heatley SL et al. Polymorphism in human cytomegalovirus UL40 impacts on recognition of human leukocyte antigen-E (HLA-E) by natural killer cells. The Journal of Biological Chemistry. 2013;**288**(12):8679-8690

[29] Garcia P et al. Human T cell receptor-mediated recognition of HLA-E. European Journal of Immunology. 2002;**32**(4):936-944

[30] Salerno-Goncalves R et al. Identification of a human HLA-E-restricted CD8+ T cell subset in volunteers immunized with Salmonella enterica serovar Typhi strain Ty21a typhoid vaccine. Journal of Immunology. 2004;**173**(9):5852-5862

[31] Riddell SR et al. Restoration of viral immunity in immunodeficient humans by the adoptive transfer of T cell clones. Science. 1992;**257**(5067):238-241

[32] Cobbold M et al. Adoptive transfer of cytomegalovirus-specific CTL to stem cell transplant patients after selection by HLA-peptide tetramers. The Journal of Experimental Medicine. 2005;**202**(3):379-386

[33] Feuchtinger T et al. Adoptive transfer of pp65-specific T cells for the treatment of chemorefractory cytomegalovirus disease or reactivation after haploidentical and matched unrelated stem cell transplantation. Blood. 2010;**116**(20):4360-4367

[34] Schmitt A et al. Adoptive transfer and selective reconstitution of streptamer-selected cytomegalovirus-specific CD8+ T cells leads to virus clearance in patients after allogeneic peripheral blood stem cell transplantation. Transfusion. 2011;**51**(3):591-599

[35] Elkington R et al. Ex vivo profiling of CD8+–T-cell responses to human cytomegalovirus reveals broad and multispecific reactivities in healthy virus carriers. Journal of Virology. 2003;**77**(9):5226-5240

[36] Sylwester AW et al. Broadly targeted human cytomegalovirus-specific CD4+ and CD8+ T cells dominate the memory compartments of exposed subjects. The Journal of Experimental Medicine. 2005;**202**(5):673-685

[37] Ameres S et al. Presentation of an immunodominant immediate-early CD8+ T cell epitope resists human cytomegalovirus immunoevasion. PLoS Pathogens. 2013;**9**(5):e1003383

[38] Gibson L et al. Human cytomegalovirus proteins pp65 and immediate early protein 1 are common targets for CD8+ T cell responses in children with congenital or postnatal human cytomegalovirus infection. Journal of Immunology. 2004;**172**(4):2256-2264

[39] Kato R et al. Early detection of cytomegalovirus-specific cytotoxic T lymphocytes against cytomegalovirus antigenemia in human leukocyte antigen haploidentical hematopoietic stem cell transplantation. Annals of Hematology. 2015;**94**(10):1707-1715

[40] Nguyen TH et al. Cross-reactive anti-viral T cells increase prior to an episode of viral reactivation post human lung transplantation. PLoS One. 2013;**8**(2):e56042

[41] van Bockel D et al. Validation of RNA-based molecular clonotype analysis for virus-specific CD8+ T-cells in formaldehyde-fixed specimens isolated from peripheral blood. Journal of Immunological Methods. 2007;**326**(1-2):127-138

[42] Yang X et al. Structural basis for clonal diversity of the public T cell response to a dominant human cytomegalovirus epitope. The Journal of Biological Chemistry. 2015;**290**(48):29106-29119

[43] Kuzushima K et al. Efficient identification of HLA-A*2402-restricted cytomegalovirus-specific CD8(+) T-cell epitopes by a computer algorithm and an enzyme-linked immunospot assay. Blood. 2001;**98**(6):1872-1881

[44] Doubrovina E et al. Adoptive immunotherapy with unselected or EBV-specific T cells for biopsy-proven EBV+ lymphomas after allogeneic hematopoietic cell transplantation. Blood. 2012;**119**(11):2644-2656

[45] Gottschalk S et al. An Epstein-Barr virus deletion mutant associated with fatal lympho-proliferative disease unresponsive to therapy with virus-specific CTLs. Blood. 2001; **97**(4):835-843

[46] Yaciuk JC et al. Direct interrogation of viral peptides presented by the class I HLA of HIV-infected T cells. Journal of Virology. 2014;**88**(22):12992-13004

[47] Krausa P et al. Definition of peptide binding motifs amongst the HLA-A*30 allelic group. Tissue Antigens. 2000;**56**(1):10-18

[48] Macdonald WA et al. A naturally selected dimorphism within the HLA-B44 supertype alters class I structure, peptide repertoire, and T cell recognition. The Journal of Experimental Medicine. 2003;**198**(5):679-691

[49] Kruger T et al. Lessons to be learned from primary renal cell carcinomas: Novel tumor antigens and HLA ligands for immunotherapy. Cancer Immunology, Immunotherapy. 2005;**54**(9):826-836

[50] Weinzierl AO et al. Distorted relation between mRNA copy number and corresponding major histocompatibility complex ligand density on the cell surface. Molecular & Cellular Proteomics. 2007;**6**(1):102-113

[51] Kunze-Schumacher H, Blasczyk R, Bade-Doeding C. Soluble HLA technology as a strategy to evaluate the impact of HLA mismatches. Journal of Immunology Research. 2014;**2014**:246171

[52] Scull KE et al. Secreted HLA recapitulates the immunopeptidome and allows in-depth coverage of HLA a*02:01 ligands. Molecular Immunology. 2012;**51**(2):136-142

[53] Bassani-Sternberg M, Coukos G. Mass spectrometry-based antigen discovery for cancer immunotherapy. Current Opinion in Immunology. 2016;**41**:9-17

[54] Slingluff CL Jr. The present and future of peptide vaccines for cancer: Single or multiple, long or short, alone or in combination? Cancer Journal. 2011;**17**(5):343-350

[55] Li W et al. Peptide vaccine: Progress and challenges. Vaccines (Basel). 2014;**2**(3):515-536

[56] Purcell AW, Croft NP, Tscharke DC. Immunology by numbers: Quantitation of antigen presentation completes the quantitative milieu of systems immunology! Current Opinion in Immunology. 2016;**40**:88-95

[57] Caron E et al. Analysis of major histocompatibility complex (MHC) Immunopeptidomes using mass spectrometry. Molecular & Cellular Proteomics. 2015;**14**(12):3105-3117

[58] Parker KC, Bednarek MA, Coligan JE. Scheme for ranking potential HLA-A2 binding peptides based on independent binding of individual peptide side-chains. Journal of Immunology. 1994;**152**(1):163-175

[59] Doytchinova IA, Guan PP, Flower DR. EpiJen: A server for multistep T cell epitope prediction. BMC Bioinformatics. 2006;**7**:131-142

[60] Peters B et al. Examining the independent binding assumption for binding of peptide epitopes to MHC-I molecules. Bioinformatics. 2003;**19**(14):1765-1772

[61] Vita R et al. The immune epitope database (IEDB) 3.0. Nucleic Acids Research. 2015; **43**(Database issue):D405-D412

[62] Kim Y et al. Immune epitope database analysis resource. Nucleic Acids Research. 2012; **40**(Web Server issue):W525-W530

[63] Wang P et al. A systematic assessment of MHC class II peptide binding predictions and evaluation of a consensus approach. PLoS Computational Biology. 2008;**4**(4):e1000048

[64] Moutaftsi M et al. A consensus epitope prediction approach identifies the breadth of murine T(CD8+)-cell responses to vaccinia virus. Nature Biotechnology. 2006;**24**(7):817-819

[65] Hakenberg J et al. MAPPP: MHC class I antigenic peptide processing prediction. Applied Bioinformatics. 2003;**2**(3):155-158

[66] Xu Y et al. MHC2MIL: A novel multiple instance learning based method for MHC-II peptide binding prediction by considering peptide flanking region and residue positions. BMC Genomics. 2014;**15**(Suppl 9):S9

[67] Bhasin M, Raghava GP. SVM based method for predicting HLA-DRB1*0401 binding peptides in an antigen sequence. Bioinformatics. 2004;**20**(3):421-423

[68] Lata S, Bhasin M, Raghava GP. MHCBN 4.0: A database of MHC/TAP binding peptides and T-cell epitopes. BMC Research Notes. 2009;**2**:61

[69] Bhasin M, Singh H, Raghava GP. MHCBN: A comprehensive database of MHC binding and non-binding peptides. Bioinformatics. 2003;**19**(5):665-666

[70] EL-Manzalawy Y, Dobbs D, Honavar V. Predicting MHC-II binding affinity using multiple instance regression. IEEE-ACM Transactions on Computational Biology and Bioinformatics. 2011;**8**(4):1067-1079

[71] Guan P et al. MHCPred: A server for quantitative prediction of peptide-MHC binding. Nucleic Acids Research. 2003;**31**(13):3621-3624

[72] Guan P et al. MHCPred: Bringing a quantitative dimension to the online prediction of MHC binding. Applied Bioinformatics. 2003;**2**(1):63-66

[73] Hattotuwagama CK et al. Quantitative online prediction of peptide binding to the major histocompatibility complex. Journal of Molecular Graphics & Modelling. 2004; **22**(3):195-207

[74] Bhasin M, Raghava GP. Prediction of promiscuous and high-affinity mutated MHC binders. Hybridoma and Hybridomics. 2003;**22**(4):229-234

[75] Zhang GL et al. MULTIPRED2: A computational system for large-scale identification of peptides predicted to bind to HLA supertypes and alleles. Journal of Immunological Methods. 2011;**374**(1-2):53-61

[76] Larsen MV et al. Large-scale validation of methods for cytotoxic T-lymphocyte epitope prediction. BMC Bioinformatics. 2007;**8**:424

[77] Andreatta M, Nielsen M. Gapped sequence alignment using artificial neural networks: Application to the MHC class I system. Bioinformatics. 2016;**32**(4):511-517

[78] Karosiene E et al. NetMHCcons: A consensus method for the major histocompatibility complex class I predictions. Immunogenetics. 2012;**64**(3):177-186

[79] Andreatta M et al. NNAlign: A web-based prediction method allowing non-expert end-user discovery of sequence motifs in quantitative peptide data. PLoS One. 2011;**6**(11): e26781

[80] Lundegaard C, Lund O, Nielsen M. Accurate approximation method for prediction of class I MHC affinities for peptides of length 8, 10 and 11 using prediction tools trained on 9mers. Bioinformatics. 2008;**24**(11):1397-1398

[81] Karosiene E et al. NetMHCIIpan-3.0, a common pan-specific MHC class II prediction method including all three human MHC class II isotypes, HLA-DR, HLA-DP and HLA-DQ. Immunogenetics. 2013;**65**(10):711-724

[82] Andreatta M et al. Accurate pan-specific prediction of peptide-MHC class II binding affinity with improved binding core identification. Immunogenetics. 2015;**67**(11-12):641-650

[83] Hoof I et al. NetMHCpan, a method for MHC class I binding prediction beyond humans. Immunogenetics. 2009;**61**(1):1-13

[84] Jurtz V et al. NetMHCpan-4.0: Improved peptide-MHC class I interaction predictions integrating eluted ligand and peptide binding affinity data. Journal of Immunology. 2017; **199**(9):3360-3368

[85] Bhasin M, Raghava GP. A hybrid approach for predicting promiscuous MHC class I restricted T cell epitopes. Journal of Biosciences. 2007;**32**(1):31-42

[86] Schueler-Furman O et al. Structure-based prediction of binding peptides to MHC class I molecules: Application to a broad range of MHC alleles. Protein Science. 2000;**9**(9): 1838-1846

[87] Singh H, Raghava GP. ProPred: Prediction of HLA-DR binding sites. Bioinformatics. 2001; **17**(12):1236-1237

[88] Singh H, Raghava GP. ProPred1: Prediction of promiscuous MHC class-I binding sites. Bioinformatics. 2003;**19**(8):1009-1014

[89] Reche PA et al. Enhancement to the RANKPEP resource for the prediction of peptide binding to MHC molecules using profiles. Immunogenetics. 2004;**56**(6):405-419

[90] Reche PA, Reinherz EL. Prediction of peptide-MHC binding using profiles. Methods in Molecular Biology. 2007;**409**:185-200

[91] Reche PA, Glutting JP, Reinherz EL. Prediction of MHC class I binding peptides using profile motifs. Human Immunology. 2002;**63**(9):701-709

[92] Donnes P, Elofsson A. Prediction of MHC class I binding peptides, using SVMHC. BMC Bioinformatics. 2002;**3**:25

[93] Donnes P, Kohlbacher O. SVMHC: A server for prediction of MHC-binding peptides. Nucleic Acids Research. 2006;**34**(Web Server issue):W194-W197

[94] Rammensee H et al. SYFPEITHI: Database for MHC ligands and peptide motifs. Immunogenetics. 1999;**50**(3-4):213-219

[95] Sturniolo T et al. Generation of tissue-specific and promiscuous HLA ligand databases using DNA microarrays and virtual HLA class II matrices. Nature Biotechnology. 1999;**17**(6):555-561

[96] Blythe MJ, Doytchinova IA, Flower DR. JenPep: A database of quantitative functional peptide data for immunology. Bioinformatics. 2002;**18**(3):434-439

[97] McSparron H et al. JenPep: A novel computational information resource for immunobiology and vaccinology. Journal of Chemical Information and Computer Sciences. 2003;**43**(4):1276-1287

[98] Wang P et al. Peptide binding predictions for HLA DR, DP and DQ molecules. BMC Bioinformatics. 2010;**11**:568

[99] Miyazawa S, Jernigan RL. Residue-residue potentials with a favorable contact pair term and an unfavorable high packing density term, for simulation and threading. Journal of Molecular Biology. 1996;**256**(3):623-644

[100] Ruppert J et al. Prominent role of secondary anchor residues in peptide binding to HLA-A2.1 molecules. Cell. 1993;**74**(5):929-937

[101] Sidney J et al. Specificity and degeneracy in peptide binding to HLA-B7-like class I molecules. Journal of Immunology. 1996;**157**(8):3480-3490

[102] Brusic V, Rudy G, Harrison LC. MHCPEP, a database of MHC-binding peptides: Update 1997. Nucleic Acids Research. 1998;**26**(1):368-371

[103] Benson DA et al. GenBank. Nucleic Acids Research. 2003;**31**(1):23-27

[104] Ahn K et al. The ER-luminal domain of the HCMV glycoprotein US6 inhibits peptide translocation by TAP. Immunity. 1997;**6**(5):613-621

[105] Park B et al. Human cytomegalovirus inhibits tapasin-dependent peptide loading and optimization of the MHC class I peptide cargo for immune evasion. Immunity. 2004;**20**(1):71-85

[106] Hill A et al. Herpes simplex virus turns off the TAP to evade host immunity. Nature. 1995;**375**(6530):411-415

[107] Vambutas A et al. Interaction of human papillomavirus type 11 E7 protein with TAP-1 results in the reduction of ATP-dependent peptide transport. Clinical Immunology. 2001; **101**(1):94-99

[108] Bennett EM et al. Cutting edge: Adenovirus E19 has two mechanisms for affecting class I MHC expression. Journal of Immunology. 1999;**162**(9):5049-5052

[109] Horst D et al. Specific targeting of the EBV lytic phase protein BNLF2a to the transporter associated with antigen processing results in impairment of HLA class I-restricted antigen presentation. Journal of Immunology. 2009;**182**(4):2313-2324

[110] Beier DC et al. Association of human class I MHC alleles with the adenovirus E3/19K protein. Journal of Immunology. 1994;**152**(8):3862-3872

[111] Kerkau T et al. The human immunodeficiency virus type 1 (HIV-1) Vpu protein interferes with an early step in the biosynthesis of major histocompatibility complex (MHC) class I molecules. The Journal of Experimental Medicine. 1997;**185**(7):1295-1305

[112] Gruhler A, Peterson PA, Fruh K. Human cytomegalovirus immediate early glycoprotein US3 retains MHC class I molecules by transient association. Traffic. 2000;**1**(4):318-325

[113] Coscoy L, Ganem D. Kaposi's sarcoma-associated herpesvirus encodes two proteins that block cell surface display of MHC class I chains by enhancing their endocytosis. Proceedings of the National Academy of Sciences of the United States of America. 2000; **97**(14):8051-8056

[114] Schwartz O et al. Endocytosis of major histocompatibility complex class I molecules is induced by the HIV-1 Nef protein. Nature Medicine. 1996;**2**(3):338-342

[115] Wiertz EJ et al. The human cytomegalovirus US11 gene product dislocates MHC class I heavy chains from the endoplasmic reticulum to the cytosol. Cell. 1996;**84**(5):769-779

[116] Wiertz EJ et al. Sec61-mediated transfer of a membrane protein from the endoplasmic reticulum to the proteasome for destruction. Nature. 1996;**384**(6608):432-438

[117] Gilbert MJ et al. Cytomegalovirus selectively blocks antigen processing and presentation of its immediate-early gene product. Nature. 1996;**383**(6602):720-722

[118] Yin Y, Manoury B, Fahraeus R. Self-inhibition of synthesis and antigen presentation by Epstein-Barr virus-encoded EBNA1. Science. 2003;**301**(5638):1371-1374

Chapter 3

The Role of Erythropoietin-Derived Peptides in Tissue Protection

Chao Zhang and Cheng Yang

Additional information is available at the end of the chapter

http://dx.doi.org/10.5772/intechopen.71931

Abstract

Erythropoietin (EPO), recognized as a tissue protective agent, can trigger anti-inflammatory and anti-apoptotic processes to delimit injury and promote repair by the binding tissue-protective receptor. However, only at a high dosage can EPO exert tissue protective effects, which simultaneously elicits some severe erythropoiesis-related side effects. Thus, the structural modification of EPO for the prevention of side effects is undoubtedly required. This chapter reviewed the development from EPO to its peptide derivatives with tissue protective efficacy. We also discussed are the therapeutic effects and limitations of each peptide, signaling pathways involved and the benefits for translation.

Keywords: erythropoietin, peptide, helix B peptide, cyclic helix B peptide, tissue protection

1. Introduction

Tissue injury refers to the histologic lesions and the subsequent functional insufficiencies of tissues and organs that are caused by multiple sources, such as ischemia followed by reperfusion [1], trauma [2], autoimmune inflammation [3], oxidative stress [4], drugs and toxicants [5], etc. In histology, it is characterized by the infiltration of pro-inflammatory cells and cytokines and the cellular necrosis and apoptosis induced by damage-associated molecular patterns [6, 7]. Considering the high morbidity and mortality it has caused, the protection from tissue injury is a major concern in the basic and clinical medical academies worldwide. Erythropoietin (EPO), recognized as the erythropoietic hormone secreted in response to anemia, was reported to trigger anti-inflammatory and anti-apoptotic processes to attenuate tissue injury and promote the repairing of injured tissues by binding to tissue protective erythropoietin receptor (EPOR)/β common receptor (βcR) heterodimer [8]. However, only at a high dosage can EPO exert tissue protective effects, which simultaneously elicits some severe erythropoiesis-related side effects [9]. Thus, the structural modification of EPO for the prevention of side effects is undoubtedly required. This chapter reviewed the development from EPO to its peptide derivatives with tissue protective efficacy. Also discussed are the therapeutic effects and limitations of each peptide, signaling pathways involved and the benefits for translation.

2. Erythropoietin: more than erythropoiesis

EPO is a glycoprotein hormone, which is predominately produced and secreted by the adult kidney in response to anemia [10, 11]. After sensing the hypoxic signal in anemic conditions, the fibroblasts in the interstitial area of the renal cortex are activated to produce EPO [12, 13]. EPO stimulates erythropoiesis through binding to its EPOR expressed on the surface of the red blood cell progenitors and precursors in bone marrow [14, 15]. In normal conditions, the level of erythropoietin in the blood is quite low [10]. Under hypoxic stress, however, EPO production may be increased up to 1000-fold, reaching 10,000 mU/ml in the blood [13]. In structure, EPO consists of four long alpha-helixes (A, B, C, and D) running in an up-up-down-down direction, connected by two long cross-over loops (AB and CD) and one short loop (BC). Among the four helixes, only helix A, C, and helix D and the loop connecting helix A and B within the dimensional structure are essential for interacting with EPOR, while helix B faces away from the interior of the receptor in the tertiary structure [13, 16–18].

During the past decade, greater interest has been paid to the pleiotropic biologic actions of EPO beyond the stimulation of erythropoiesis, which include anti-apoptosis, anti-inflammation, neurogenesis, and angiogenesis, as well as their consequent tissue protection [8, 19–21]. An increasing body of studies demonstrated that EPO exerts a powerful tissue-protective effect on a variety of organs and can prevent cellular apoptosis from some sources, including reduced or absent oxygen tension, ischemia–reperfusion, toxicity and free radical exposure. Our study in a murine model of renal injury showed that EPO protected kidneys against injury through decreasing positive myeloperoxidase neutrophils and suppressing the expression of pro-inflammatory cytokines and chemokines by the inhibition of NF-kB signaling pathway [22]. In the study with isolated porcine renal allografts, it was demonstrated that EPO promoted inflammatory cell apoptosis, drove inflammatory and apoptotic cells into tubular lumens, thereby leading to inflammation clearance in the involved tissue and organs [23]. We also reported in our recent study that EPO could affect the dynamics of macrophages in the model of rhabdomyolysis-induced acute kidney injury. In this study, EPO was found to ameliorate kidney injury by reducing macrophages recruitment and promoting phenotype switch from M1 to M2 [24]. In vitro study, EPO was shown to directly suppress pro-inflammatory responses of M1 macrophages and promote M2 marker expression [24]. These results refreshed our understanding about the immunoregulatory capacity of EPO.

The mechanisms involved in the tissue protective effect of EPO was not well illustrated until the specific receptor mediating tissue protection was revealed, which was recently defined as a heterodimer composed of EPOR and CD131 [25]. CD131 also forms the receptor complexes with α receptor subunits specific for GM-CSF, IL-3, and IL-5 and thus is named as β common receptor [25]. In aqueous media, helix B and parts of the AB and CD loops face the aqueous medium but away from the erythropoietic binding sites, which indicates that helix B mediates tissue protective effect of EPO via binding with EPOR/βcR heterodimer [26]. Following the interaction of EPO with EPOR/βcR, the Akt involved signaling pathway is reported to play a vital role in mediating intracellular signal transduction. The activation of PI3K/Akt pathway by EPO maintains the mitochondrial membrane integrity, prevents cytochrome C from release and modulates the activity of caspase cascade during cellular apoptosis [27, 28]. The blockade of Akt phosphorylation abrogates the anti-apoptotic and anti-inflammatory effect after EPO administration [28, 29]. Moreover, the activation of Akt after EPO treatment is also reported to protect against genomic DNA degradation and membrane phosphatidylserine exposure [28, 30]. Several transcriptional modulators in the downstream pathways of Akt have also been shown to mediate the tissue-protection of EPO. For example, EPO could down-regulate the activity of forkhead transcription factor (FOXO3a) through inhibitory phosphorylation, which renders FOXO3a ineffective to activate the transcription of nuclear genes involved in apoptosis [31]. Other mechanisms include the inhibition of glycogen synthase kinase-3β [32], a serine-threonine kinase that plays a significant role in the induction of apoptosis of neurons, vascular smooth muscle cells and cardiomyocytes, and the up-regulation of the anti-apoptotic Bcl-2 family member Bcl-xL [33].

3. Helix B surface peptide: a specific tissue protective peptide without erythropoietic effect

The role of EPO in anti-inflammation and anti-apoptosis inspired us to investigate if EPO could also be served as a potential therapy for tissue injury caused by the exposure to drugs, chemicals, or physical ischemia. However, the tissue protective effect of EPO only occurs at the dosage that is well above normal, which may simultaneously elicit severe side-effects associated with its erythropoietic effects such as polycythemia and hypercoagulation in the circulation [9]. The demand of high dose to exert the tissue protective activities may be caused by the relatively low affinity of EPO to the tissue-protective EPOR/βcR dimer receptor relative to the erythropoietic (EPOR)$_2$ homodimer [15, 20, 34]. Being enlightened by the finding that helix B plays a dominant role in mediating the tissue protective effect by binding with EPOR/βcR receptor, Michael Brines et al. first synthesized the nonerythropoietic, tissue-protective peptides derived from the tertiary structure of erythropoietin which comprises the amino acid sequence corresponding to helix B as well as the three residues within the proximal portion of the BC loop. The design of the novel peptide aims to mimic the three-dimensional structure that interacts with EPOR/βcR receptor to reproduce the tissue-protective activities of the full molecule [26]. This 11-mer peptide derivative was named as helix B surface peptide (HBSP) [26]. Following studies demonstrated that HBSP is sufficient to activate tissue-protective pathways representative of the full molecule and protect from injury in a wide variety of tissues and organs but without causing erythropoietic effects.

3.1. In cardiovascular system

HBSP was demonstrated to exert tissue protective effects first in the cardiovascular system after it was designed. The designers of this peptide showed that HBSP protected cardiomyocytes from TNF-α apoptosis in an Akt-dependent pathway both in vitro and in vivo [26, 35]. In this study, the levels of serum creatinine kinase activity and of cardiac expression of atrial natriuretic peptide, a marker of chronic heart failure, were down-regulated in animals treated with HBSP [26, 35]. Then, the anti-atherosclerotic effects HBSP were investigated in vitro and in vivo [36]. In vitro, HBSP inhibited C-reactive protein induced apoptosis in human umbilical vein endothelial cells and THP-1 cells to a great extent [36]. In the hyperlipidemic spontaneous myocardial infarction model of rabbits, HBSP was shown to significantly suppress the progression of coronary stenosis and myocardial ischemia caused by atherosclerotic lesions and inhibit coronary artery endothelial cell apoptosis through the activation of Akt pathway and the corresponding decreased the production of TNF-α as well as modified macrophage M1/M2 polarization [36]. Similarly, in the myocardial ischemia-reperfusion injury model of mice, HBSP administration before reperfusion significantly reduced the myocardial infarct size, decreased cardiomyocyte apoptosis, reduced the activities of superoxide dismutase and partially preserved heart function through the upregulation of Akt/GSK-3β/ERK and STAT-3 [37]. In an in vitro study performed by using a rodent cardiomyocyte cell line subjected to hypoxia-reoxygenation injury, HBSP was reported to have protective effects by reducing cellular apoptosis, mitochondrial reactive oxygen production, $\Delta\Psi m$ collapse, and cytochrome C release from mitochondria to the cytosol. Furthermore, HBSP inhibited the activation of caspase 9 and caspase 3 as well as the alteration of Bcl-2 family proteins induced by hypoxia-reoxygenation [38]. Diabetes is also one of the major causes of myocardial lesions. Diabetic cardiomyopathy (DCM) is a ventricular dysfunction independent of coronary artery disease and hypertension, which is associated with inflammation, myocardial apoptosis and fibrosis [39–41]. In a study regarding the protection of HBSP on DCM, HBSP notably improved cardiac function, attenuated cardiac interstitial fibrosis, inhibited myocardial apoptosis, and ameliorated mitochondrial ultrastructure in mice with diabetic cardiomyopathy through an AMPK-dependent pathway [42]. HBSP promoted aortic endothelial cell repair under hypoxic conditions in a model of aortic endothelial injury, in which HBSP enhanced scratch closure by promoting cell migration and proliferation [43]. Furthermore, EPO protected bovine aortic endothelial cells from staurosporine-induced apoptosis under hypoxic conditions. Hypoxia was associated with a reduction in nitric oxide (NO) production. HBSP notably increased NO production, in a manner sensitive to NO synthase inhibition, under hypoxic conditions but not under normoxic conditions [43]. In summary, multiple studies proved the protective effect of HBSP on myocardial tissues and endothelial cells in the cardiovascular system following the injury caused by insufficient oxygen supply and implied that Akt pathway played a critical role in this process.

3.2. In nervous system

Another field of research in which the protection of HBSP was well investigated is in the nervous system. In one study by Robertson et al., the effects of HBSP on early cerebral hemodynamics and neurological outcome post-injury were investigated in a rat model of mild cortical impact injury followed hemorrhagic hypotension [44]. The results demonstrated that both EPO and HBSP treated groups improved recovery of cerebral blood flow in the injured brain following resuscitation, and showed more rapid recovery in the performance of functional neural tests. This study suggests that HBSP has neuroprotective effects similar to EPO in this model of combined brain injury and hypotension [44]. They later reported that the treatment with HBSP resulted in significantly improved performance after the rats were suffered from mild traumatic brain injury (mTBI), which was associated with decreased infiltration of CD68-positive inflammatory cells in the damaged brain tissue [45]. The results suggest that HBSP may improve cognitive function following mTBI [45]. Among the patients with neuritis, neuropathic pain is a quite common symptom, which may be due to nerve inflammation. The neuropathic pain results from several overlapping pathways, which then merges into a magnified pain status with symptoms such as allodynia and hyperalgesia [46, 47]. The study by Pulman KG et al., examined the effects of HBSP on pain behavior in the rat model of neuritis [48]. The results showed that treatment with HBSP prevented the development of mechanical allodynia caused by neuritis but not affect heat hyperalgesia [48]. In another study, HBSP treatment could reduce allodynia coupled to the suppression of spinal microglia response in a dose-dependent manner, which may result from HBSP-induced suppression of inflammation in central nervous system [49]. In the study on the therapeutic effects of HBSP in experimental autoimmune encephalomyelitis (EAE), the administration of HBSP to EAE rats significantly reduced the severity and shortened the duration of injury, reduced the infiltration of pro-inflammatory cells and suppressed expression of pro-inflammatory cytokines such as IL-1β, IL-17, TNF-α, IFN-γ. The expression of inducible NO synthase and transcription factor T-bet at mRNA level was also reduced in spinal cords following HBSP treatment [50]. In the in vitro study, HBSP inhibited antigen-specific and non-specific lymphocyte proliferation and promoted the polarization of Th2 and regulatory T cells (Treg) while suppressed the polarization of Th1 and Th17 cells in EAE lymph nodes [50]. In summary, these studies regarding the role of HBSP in nervous system revealed that HBSP could also protect the nervous tissue from injury, which was associated with the alteration of the cytokine and cell milieu to limit inflammation in the damaged nervous tissue.

3.3. In obesity and diabetes related disorders

Several recent studies focused on the evaluation of the effects and potential mechanisms of HBSP in obesity modulation and diabetes-related disorders. One study was performed by using male C57BL/6 J mice fed with high-fat high-sucrose (HFHS) diet [51]. HFHS diet treated mice exhibited insulin resistance, hyperlipidemia, hepatic lipid accumulation and kidney dysfunction, which was related to the impaired insulin signal pathway and reduced membrane translocation of glucose transporter 4 [51]. However, treatment with HBSP ameliorated renal function, reduced hepatic lipid deposition, and normalized serum glucose and lipid profiles, which were associated with improved insulin sensitivity and glucose uptake in skeletal muscle. The mechanism included that HBSP attenuated the HFHS-induced overproduction of IL-6 and fibroblast growth factor-21, and enhanced mitochondrial biogenesis in skeletal muscle [51]. In another study regarding the effect of HBSP on obesity, HBSP was found to protect against obesity and insulin resistance by suppressing adipogenesis, adipokine expression as well as attenuating macrophage inflammatory activation in lipid tissue [52]. The retinopathy is one of the most common complications of diabetes and remains one of the leading causes of non-congenital blindness [53]. In the diabetic retina, vasodegenerative phase is accompanied by neuroglial abnormalities and eventual depletion of ganglion cells [54]. HBSP was shown to significantly reduce microglial activation and protected against neuroglial and vascular degeneration but without exacerbating neovascularization in the retina [55]. These findings suggest that HBSP has therapeutic implications for metabolic disorders, such as obesity, diabetes, and diabetic retinopathy.

3.4. In kidney

Our research mainly discussed the tissue protection of HBSP on kidney injury. In 2013, we investigated effects of HBSP and the expression of EPOR/βcR heterodimer receptor in a murine renal ischemia-reperfusion (IR) injury model [56]. We found that HBSP could significantly ameliorate renal dysfunction and tissue damage, reduced apoptotic cells in the kidney and inhibited the activation of caspase-9 and caspase-3 [56]. The expression of EPOR/βcR in the kidney was up-regulated following ischemia-reperfusion injury but was down-regulated by the treatment of HBSP [56]. Further investigation revealed that the PI3K-Akt pathway was dramatically activated by HBSP. The treatment of the PI3K inhibitor, Wortmannin, abolished improved renal function and histologic structure by HBSP [56]. This study suggests that HBSP could protect the kidney from IR injury in a PI3K-Akt dependent pathway. Then, we also investigated the role of HBSP in IR and cyclosporine A (CsA) induced kidney injury since both of them are unavoidable after kidney transplantation and associated with allograft dysfunction [57]. We found that the level of creatinine and blood urea nitrogen was increased by CsA but decreased by HBSP. HBSP also significantly ameliorated tubulointerstitial damage and interstitial fibrosis, which were gradually increased by IR and CsA [57]. In addition, apoptotic cells, infiltrated inflammatory cells, and active caspase-3 positive cells were greatly reduced by HBSP. It was demonstrated for the first time that HBSP effectively improved renal function and tissue damage caused by IR and/or CsA, which might be through reducing caspase-3 activation and synthesis, apoptosis, and inflammation [57]. Similar findings were also reported by Nimesh SA Patel's and Willem G van Rijt's groups that HBSP has renoprotective capacities by anti-inflammation and anti-apoptosis in the injured kidney tissue [58, 59].

3.5. In liver

Very recently, the protective effect of HBSP on acute liver injury was investigated in Wu's study. In this study, the acute liver injury was induced by the administration of carbon tetrachloride (CCl4) [60]. HBSP was demonstrated to significantly decrease serum alanine aminotransferase, aspartate aminotransferase, lactate dehydrogenase, and pro-inflammatory cytokines in liver tissues after CCl4 injection. The infiltration of CD3, CD8, and CD68 positive cells and the expression of cleaved caspase-3 were also significantly decreased by HBSP treatment. The glutathione peroxidase activity and survival rate increased, while the total apoptotic rate was reduced in the HBSP-treated group. As to the mechanism, the authors reported that HBSP activated the PI3K/Akt/mTORC1 pathway [60]. Thus, HBSP showed convincing protective effects on CCl4-induced acute liver injury by ameliorating inflammation and apoptosis [60].

4. Thioether-cyclized helix B peptide: a cyclized peptide with increased instability and tissue protective potency

Despite the powerful tissue-protective function exhibited in various organs by inhibiting inflammation and apoptosis, the property of poor permeability to biomembranes, unstable secondary structure, and short half-time restricts the application of HBSP in translation study [26]. Therefore, the structurally optimized transformation of HBSP is urgently required. It is acknowledged that peptide cyclization could provide an efficient strategy to overcome these problems [61]. Provoked by this, we for the first time introduced the head-to-tail cyclization to the structure of HBSP to improve its stability, since the backbone of the peptide was constrained by the cyclization in which the linkages between main chains were formed by thioether [62]. This newly designed and synthesized peptide was named as thioether-cyclized helix B peptide (CHBP) [62].

4.1. Novel properties and mechanisms

In the following study, we demonstrate that CHBP is significantly stable in the human plasma and has a 2.5-fold longer half-life time than HBSP, suggesting that CHBP is highly resistant to proteolytic degradation both in vitro and in vivo [63]. We also found in our study that due to its stability, this long-acting peptide could ameliorate renal IR injury to a greater extent than HBSP, for only one dose of CHBP exerted persistent renal protective effect throughout the one week post IR injury [63]. Autophagy is demonstrated to play a renoprotective role in IR injury and is closely related to cellular apoptosis and inflammation in kidney tissue [64]. We also found that CHBP could induce autophagy in the injured kidney by increasing LC3-II/I ratio as well as upregulating beclin-1 [62]. Furthermore, our study depicted possible signaling pathways involved in CHBP-induced autophagy, which included the regulation of mammalian target of rapamycin (mTOR) pathway and the activation of AMPK pathway. The activation of AMPK by CHBP then phosphorylated and activated tuberous sclerosis 2 (TSC2), which connected with tuberous sclerosis 1(TSC1) to form a heterodimer to inhibit the activation of mammalian target of rapamycin complex 1 (mTORC1). Meanwhile, the mTORC2-Akt pathway was activated by CHBP and autophagy was induced by the altered mTORC1/mTORC2 equilibrium [62]. Also, CHBP was reported to upregulate Treg and downregulate helper T cell 17 (Th17) after renal IR injury to restore the Treg/Th17 balance [62]. These findings revealed the mechanisms that are involved in the tissue protective function in CHBP but have not been reported in HBSP yet.

4.2. CHBP in organ preservation

During the transport of donated organs, any strategies that can effectively protect against IR injury during the cold storage (CS) and reperfusion stages would be very beneficial for preventing the delayed graft function after kidney transplant surgery. Thus, we administrated CHBP in the preservation solution and autologous blood perfusate to examine its effect on the preservation of isolated donor kidney in the following study [65]. The results showed that the administration of CHBP during cold preservation of kidneys as well as autologous blood could ameliorate IR injury after hemoperfusion, which was associated with increased renal blood supply and improved renal tubular structure and function [65].

4.3. CHBP and anti-allograft rejection

As the professional antigen-presenting cells, dendritic cells (DCs) play a triggering role in acute rejection (AR) after transplant surgery. Thus, we investigated the effects of CHBP on DCs in the kidney transplantation model from Lewis to Wistar rats [66]. The results showed that five successive doses of CHBP administration after kidney transplantation could significantly ameliorate AR with the association of lower histological injury, apoptosis, and CD4+ and CD8+ T-cell infiltration in renal allografts. CHBP also reduced the expression of IFN-γ and IL-1β but increased the expression of IL-4 and IL-10 in the serum of receipt. The number of mature DCs was significantly decreased in renal allografts treated with CHBP [66]. Also, the incubation of DCs with CHBP in vitro led to a reduction in TNF-α, IFN-γ, IL-1β and IL-12 levels and an increase of IL-10 level at the protein level in the supernatant [66]. In the mechanism study, CHBP inhibited TLR activation-induced DC maturation by increasing SOCS1 expression through Jak-2/STAT3 signaling [66]. Our study suggested that CHBP suppressed renal allograft AR by inhibiting the maturation of DCs via Jak-2/STAT3/SOCS1 signaling.

4.4. CHBP and protection of mesenchymal stem cells

Mesenchymal stem cell (MSC) is a pluripotent stem cell originating from the mesoderm and has the potential to differentiate into multiple types of cells and tissues [67, 68]. Thus, MSC has long been considered as an ideal cell-based therapy in the repairing of tissue injuries. After adoptive transferred in vivo, however, MSCs may confront a variety of undesirable factors that could decrease their viability and activity [69, 70]. Among them, nutrient starvation is the major obstacle for MSCs within injured tissues. In our study regarding the effect of CHBP on MSCs in vitro, we found that CHBP could significantly improve the cell viability and suppress apoptosis of MSCs in a dose-dependent manner [71]. Starvation resulted in the mitochondrial dysfunction, and the treatment of CHBP could alleviate mitochondrial dysfunction by diminishing the oxidative stress from ROS, restore mitochondrial membrane potential and maintain mitochondrial membrane integrity through the activation of Nrf2/Sirt3/FoxO3a pathway [71]. Moreover, MSCs pretreated with CHBP were more resistant to nutrient starvation [71]. This study suggests that CHBP has the e prospects for sustaining stem cell survival under nutrient-deprived conditions and improving the therapeutic effect of MSC-based treatment.

5. Perspective and limitation for translation into clinic

The research about CHBP in our center also includes its effects on other kinds of injuries, for example, aristolochic acid-induced acute kidney injury [72]. The study on the role of CHBP in acute and chronic allograft rejection is in progress as well. Although our understanding about CHBP has significantly increased, there is still plenty of work to do to translate this protective peptide into clinical practice. For example, the pharmacokinetics and pharmacodynamics of CHBP are not examined so far. The dosage form design of this new drug should be improved for oral administration or intravenous injection. The clinical trials are indispensable before it is finally applied for clinical use. In a further study, we plan to investigate the effects of CHBP in primate models of acute organ injury which could better represent the analogous disorders in the human being. We believe that this smaller but stronger peptide derivative of EPO could facilitate the treatment of acute tissue injury shortly.

Acknowledgements

This study was supported by National Natural Science Foundation of China (grants 81400752; 81770746 to CY).

Conflict of interest

The authors declare no conflict of interest.

Author details

Chao Zhang[1,2] and Cheng Yang[1,3]*

*Address all correspondence to: esuperyc@163.com

1 Department of Urology, Zhongshan Hospital, Fudan University, Shanghai, China

2 Department of Pharmacology and Toxicology, University of Mississippi Medical Center, Jackson, United States

3 Shanghai Key Laboratory of Organ Transplantation, Shanghai, China

References

[1] Carden DL, Granger DN. Pathophysiology of ischaemia-reperfusion injury. The Journal of Pathology. 2000;**190**(3):255-266

[2] Neher MD, Weckbach S, Flierl MA, Huber-Lang MS, Stahel PF. Molecular mechanisms of inflammation and tissue injury after major trauma--is complement the "bad guy"? Journal of Biomedical Science. 2011;**18**:90

[3] Laskin DL, Pendino KJ. Macrophages and inflammatory mediators in tissue injury. Annual Review of Pharmacology and Toxicology. 1995;**35**:655-677

[4] Slater TF. Free-radical mechanisms in tissue injury. The Biochemical Journal. 1984;**222**(1): 1-15

[5] Choudhury D, Ahmed Z. Drug-associated renal dysfunction and injury. Nature Clinical Practice. Nephrology. 2006;**2**(2):80-91

[6] Ward PA, Warren JS, Johnson KJ. Oxygen radicals, inflammation, and tissue injury. Free Radical Biology & Medicine. 1988;**5**(5-6):403-408

[7] Dallegri F, Ottonello L. Tissue injury in neutrophilic inflammation. Inflammation Research. 1997;**46**(10):382-391   ·

[8] Brines M, Cerami A. Discovering erythropoietin's extra-hematopoietic functions: Biology and clinical promise. Kidney International. 2006;**70**(2):246-250

[9] Krapf R, Hulter HN. Arterial hypertension induced by erythropoietin and erythropoiesis-stimulating agents (ESA). Clinical Journal of the American Society of Nephrology. 2009; **4**(2):470-480

[10] Krantz SB. Erythropoietin. Blood. 1991;**77**(3):419-434

[11] Goodnough LT, Price TH, Parvin CA. The endogenous erythropoietin response and the erythropoietic response to blood loss anemia: The effects of age and gender. The Journal of Laboratory and Clinical Medicine. 1995;**126**(1):57-64

[12] Eckardt KU, Boutellier U, Kurtz A, Schopen M, Koller EA, Bauer C. Rate of erythropoietin formation in humans in response to acute hypobaric hypoxia. Journal of Applied Physiology (1985). 1989;**66**(4):1785-1788

[13] Jelkmann W. Erythropoietin: Structure, control of production, and function. Physiological Reviews. 1992;**72**(2):449-489

[14] Fraser JK, Lin FK, Berridge MV. Expression of high affinity receptors for erythropoietin on human bone marrow cells and on the human erythroleukemic cell line, HEL. Experimental Hematology. 1988;**16**(10):836-842

[15] Lee R, Kertesz N, Joseph SB, Jegalian A, Wu H. Erythropoietin (Epo) and EpoR expression and 2 waves of erythropoiesis. Blood. 2001;**98**(5):1408-1415

[16] Wen D, Boissel JP, Tracy TE, Gruninger RH, Mulcahy LS, Czelusniak J, et al. Erythropoietin structure-function relationships: High degree of sequence homology among mammals. Blood. 1993;**82**(5):1507-1516

[17] Lappin TR, Winter PC, Elder GE, McHale CM, Hodges VH, Bridges JM. Structure-function relationships of the erythropoietin molecule. Annals of the New York Academy of Sciences. 1994;**718**:191-201; discussion-2

[18] Boissel JP, Lee WR, Presnell SR, Cohen FE, Bunn HF. Erythropoietin structure-function relationships. Mutant proteins that test a model of tertiary structure. The Journal of Biological Chemistry. 1993;**268**(21):15983-15993

[19] Siren AL, Fratelli M, Brines M, Goemans C, Casagrande S, Lewczuk P, et al. Erythropoietin prevents neuronal apoptosis after cerebral ischemia and metabolic stress. Proceedings of the National Academy of Sciences of the United States of America. 2001;**98**(7):4044-4049

[20] Leist M, Ghezzi P, Grasso G, Bianchi R, Villa P, Fratelli M, et al. Derivatives of erythropoietin that are tissue protective but not erythropoietic. Science. 2004;**305**(5681):239-242

[21] Ghezzi P, Brines M. Erythropoietin as an antiapoptotic, tissue-protective cytokine. Cell Death and Differentiation. 2004;**11**(Suppl 1):S37-S44

[22] Hu L, Yang C, Zhao T, Xu M, Tang Q, Yang B, et al. Erythropoietin ameliorates renal ischemia and reperfusion injury via inhibiting tubulointerstitial inflammation. The Journal of Surgical Research. 2012;**176**(1):260-266

[23] Yang B, Hosgood SA, Bagul A, Waller HL, Nicholson ML. Erythropoietin regulates apoptosis, inflammation and tissue remodelling via caspase-3 and IL-1beta in isolated hemoperfused kidneys. European Journal of Pharmacology. 2011;**660**(2-3):420-430

[24] Wang S, Zhang C, Li J, Niyazi S, Zheng L, Xu M, et al. Erythropoietin protects against rhabdomyolysis-induced acute kidney injury by modulating macrophage polarization. Cell Death & Disease. 2017;**8**(4):e2725

[25] Brines M, Grasso G, Fiordaliso F, Sfacteria A, Ghezzi P, Fratelli M, et al. Erythropoietin mediates tissue protection through an erythropoietin and common beta-subunit heteroreceptor. Proceedings of the National Academy of Sciences of the United States of America. 2004;**101**(41):14907-14912

[26] Brines M, Patel NS, Villa P, Brines C, Mennini T, De Paola M, et al. Nonerythropoietic, tissue-protective peptides derived from the tertiary structure of erythropoietin. Proceedings of the National Academy of Sciences of the United States of America. 2008;**105**(31):10925-10930

[27] Tramontano AF, Muniyappa R, Black AD, Blendea MC, Cohen I, Deng L, et al. Erythropoietin protects cardiac myocytes from hypoxia-induced apoptosis through an Akt-dependent pathway. Biochemical and Biophysical Research Communications. 2003;**308**(4): 990-994

[28] Chong ZZ, Kang JQ, Maiese K. Erythropoietin is a novel vascular protectant through activation of Akt1 and mitochondrial modulation of cysteine proteases. Circulation. 2002;**106**(23):2973-9

[29] Chong ZZ, Lin SH, Kang JQ, Maiese K. Erythropoietin prevents early and late neuronal demise through modulation of Akt1 and induction of caspase 1, 3, and 8. Journal of Neuroscience Research. 2003;**71**(5):659-69

[30] Kang JQ, Chong ZZ, Maiese K. Critical role for Akt1 in the modulation of apoptotic phosphatidylserine exposure and microglial activation. Molecular Pharmacology. 2003; **64**(3):557-569

[31] Mahmud DL, GA M, Deb DK, Platanias LC, Uddin S, Wickrema A. Phosphorylation of forkhead transcription factors by erythropoietin and stem cell factor prevents

acetylation and their interaction with coactivator p300 in erythroid progenitor cells. Oncogene. 2002;**21**(10):1556-1562

[32] Shingo T, Sorokan ST, Shimazaki T, Weiss S. Erythropoietin regulates the in vitro and in vivo production of neuronal progenitors by mammalian forebrain neural stem cells. The Journal of Neuroscience. 2001;**21**(24):9733-9743

[33] Chong ZZ, Kang JQ, Maiese K. Apaf-1, Bcl-xL, cytochrome c, and caspase-9 form the critical elements for cerebral vascular protection by erythropoietin. Journal of Cerebral Blood Flow and Metabolism. 2003;**23**(3):320-330

[34] Syed RS, Reid SW, Li C, Cheetham JC, Aoki KH, Liu B, et al. Efficiency of signalling through cytokine receptors depends critically on receptor orientation. Nature. 1998;**395**(6701):511-516

[35] Ueba H, Brines M, Yamin M, Umemoto T, Ako J, Momomura S, et al. Cardioprotection by a nonerythropoietic, tissue-protective peptide mimicking the 3D structure of erythropoietin. Proceedings of the National Academy of Sciences of the United States of America. 2010;**107**(32):14357-14362

[36] Ueba H, Shiomi M, Brines M, Yamin M, Kobayashi T, Ako J, et al. Suppression of coronary atherosclerosis by helix B surface peptide, a nonerythropoietic, tissue-protective compound derived from erythropoietin. Molecular Medicine. 2013;**19**:195-202

[37] Liu P, You W, Lin L, Lin Y, Tang X, Liu Y, et al. Helix B surface peptide protects against acute myocardial ischemia-reperfusion injury via the RISK and SAFE pathways in a mouse model. Cardiology. 2016;**134**(2):109-117

[38] Liu P, Lin Y, Tang X, Zhang P, Liu B, Liu Y, et al. Helix B surface peptide protects cardiomyocytes against hypoxia/reoxygenation-induced apoptosis through mitochondrial pathways. Journal of Cardiovascular Pharmacology. 2016;**67**(5):418-426

[39] Kannel WB, McGee DL. Diabetes and cardiovascular disease. The Framingham study. Journal of the American Medical Association. 1979;**241**(19):2035-2038

[40] Asbun J, Villarreal FJ. The pathogenesis of myocardial fibrosis in the setting of diabetic cardiomyopathy. Journal of the American College of Cardiology. 2006;**47**(4):693-700

[41] Mano Y, Anzai T, Kaneko H, Nagatomo Y, Nagai T, Anzai A, et al. Overexpression of human C-reactive protein exacerbates left ventricular remodeling in diabetic cardiomyopathy. Circulation Journal. 2011;**75**(7):1717-1727

[42] Lin C, Zhang M, Zhang Y, Yang K, Hu J, Si R, et al. Helix B surface peptide attenuates diabetic cardiomyopathy via AMPK-dependent autophagy. Biochemical and Biophysical Research Communications. 2017;**482**(4):665-671

[43] Heikal L, Ghezzi P, Mengozzi M, Stelmaszczuk B, Feelisch M, Ferns GA. Erythropoietin and a nonerythropoietic peptide analog promote aortic endothelial cell repair under hypoxic conditions: Role of nitric oxide. Hypoxia (Auckland). 2016;**4**:121-133

[44] Robertson CS, Cherian L, Shah M, Garcia R, Navarro JC, Grill RJ, et al. Neuroprotection with an erythropoietin mimetic peptide (pHBSP) in a model of mild traumatic brain injury complicated by hemorrhagic shock. Journal of Neurotrauma. 2012;**29**(6):1156-1166

[45] Robertson CS, Garcia R, Gaddam SS, Grill RJ, Cerami Hand C, Tian TS, et al. Treatment of mild traumatic brain injury with an erythropoietin-mimetic peptide. Journal of Neurotrauma. 2013;**30**(9):765-774

[46] Baron R, Binder A, Wasner G. Neuropathic pain: Diagnosis, pathophysiological mechanisms, and treatment. Lancet Neurology. 2010;**9**(8):807-819

[47] Costigan M, Scholz J, Woolf CJ. Neuropathic pain: A maladaptive response of the nervous system to damage. Annual Review of Neuroscience. 2009;**32**:1-32

[48] Pulman KG, Smith M, Mengozzi M, Ghezzi P, Dilley A. The erythropoietin-derived peptide ARA290 reverses mechanical allodynia in the neuritis model. Neuroscience. 2013;**233**:174-183

[49] Swartjes M, van Velzen M, Niesters M, Aarts L, Brines M, Dunne A, et al. ARA 290, a peptide derived from the tertiary structure of erythropoietin, produces long-term relief of neuropathic pain coupled with suppression of the spinal microglia response. Molecular Pain. 2014;**10**:13

[50] Chen H, Luo B, Yang X, Xiong J, Liu Z, Jiang M, et al. Therapeutic effects of nonerythropoietic erythropoietin analog ARA290 in experimental autoimmune encephalomyelitis rat. Journal of Neuroimmunology. 2014;**268**(1-2):64-70

[51] Collino M, Benetti E, Rogazzo M, Chiazza F, Mastrocola R, Nigro D, et al. A non-erythropoietic peptide derivative of erythropoietin decreases susceptibility to diet-induced insulin resistance in mice. British Journal of Pharmacology. 2014;**171**(24):5802-5815

[52] Liu Y, Luo B, Shi R, Wang J, Liu Z, Liu W, et al. Nonerythropoietic erythropoietin-derived peptide suppresses Adipogenesis, inflammation, obesity and insulin resistance. Scientific Reports. 2015;**5**:15134

[53] Roodhooft JM. Leading causes of blindness worldwide. Bulletin de la Société Belge d' Ophtalmologie. 2002;**283**:19-25

[54] Gardner TW, Antonetti DA, Barber AJ, LaNoue KF, Nakamura M. New insights into the pathophysiology of diabetic retinopathy: Potential cell-specific therapeutic targets. Diabetes Technology & Therapeutics. 2000;**2**(4):601-608

[55] McVicar CM, Hamilton R, Colhoun LM, Gardiner TA, Brines M, Cerami A, et al. Intervention with an erythropoietin-derived peptide protects against neuroglial and vascular degeneration during diabetic retinopathy. Diabetes. 2011;**60**(11):2995-3005

[56] Yang C, Zhao T, Lin M, Zhao Z, Hu L, Jia Y, et al. Helix B surface peptide administered after insult of ischemia reperfusion improved renal function, structure and apoptosis through beta common receptor/erythropoietin receptor and PI3K/Akt pathway in a murine model. Experimental Biology and Medicine (Maywood, N.J.). 2013;**238**(1):111-119

[57] Wu Y, Zhang J, Liu F, Yang C, Zhang Y, Liu A, et al. Protective effects of HBSP on ischemia reperfusion and cyclosporine a induced renal injury. Clinical & Developmental Immunology. 2013;**2013**:758159

[58] van Rijt WG, Nieuwenhuijs-Moeke GJ, van Goor H, Ottens PJ, Ploeg RJ, Leuvenink HG. Renoprotective capacities of non-erythropoietic EPO derivative, ARA290, following renal ischemia/reperfusion injury. Journal of Translational Medicine. 2013;**11**:286

[59] Patel NS, Kerr-Peterson HL, Brines M, Collino M, Rogazzo M, Fantozzi R, et al. Delayed administration of pyroglutamate helix B surface peptide (pHBSP), a novel nonerythropoietic analog of erythropoietin, attenuates acute kidney injury. Molecular Medicine. 2012;**18**:719-727

[60] Wu S, Yang C, Xu N, Wang L, Liu Y, Wang J, et al. The protective effects of helix B surface peptide on experimental acute liver injury induced by carbon tetrachloride. Digestive Diseases and Sciences. 2017;**62**(6):1537-1549

[61] Song YL, Peach ML, Roller PP, Qiu S, Wang S, Long YQ. Discovery of a novel nonphosphorylated pentapeptide motif displaying high affinity for Grb2-SH2 domain by the utilization of 3'-substituted tyrosine derivatives. Journal of Medicinal Chemistry. 2006;**49**(5):1585-1596

[62] Yang C, Xu Z, Zhao Z, Li L, Zhao T, Peng D, et al. A novel proteolysis-resistant cyclic helix B peptide ameliorates kidney ischemia reperfusion injury. Biochimica et Biophysica Acta. 2014;**1842**(11):2306-2317

[63] Yang C, Liu J, Li L, Hu M, Long Y, Liu X, et al. Proteome analysis of renoprotection mediated by a novel cyclic helix B peptide in acute kidney injury. Scientific Reports. 2015;**5**:18045

[64] Codogno P, Meijer AJ. Autophagy and signaling: Their role in cell survival and cell death. Cell Death and Differentiation. 2005;**12**(Suppl 2):1509-1518

[65] Yang C, Hosgood SA, Meeta P, Long Y, Zhu T, Nicholson ML, et al. Cyclic helix B peptide in preservation solution and autologous blood perfusate ameliorates ischemia-reperfusion injury in isolated porcine kidneys. Transplant Direct. 2015;**1**(2):e6

[66] Yang C, Zhang Y, Wang J, Li L, Wang L, Hu M, et al. A novel cyclic helix B peptide inhibits dendritic cell maturation during amelioration of acute kidney graft rejection through Jak-2/STAT3/SOCS1. Cell Death & Disease. 2015;**6**:e1993

[67] Bianco P, Robey PG, Simmons PJ. Mesenchymal stem cells: Revisiting history, concepts, and assays. Cell Stem Cell. 2008;**2**(4):313-319

[68] Keating A. Mesenchymal stromal cells: New directions. Cell Stem Cell. 2012;**10**(6):709-716

[69] Haider H, Ashraf M. Preconditioning and stem cell survival. Journal of Cardiovascular Translational Research. 2010;**3**(2):89-102

[70] Zhang Q, Liu S, Li T, Yuan L, Liu H, Wang X, et al. Preconditioning of bone marrow mesenchymal stem cells with hydrogen sulfide improves their therapeutic potential. Oncotarget. 2016;**7**(36):58089-58104

[71] Wang S, Zhang C, Niyazi S, Zheng L, Li J, Zhang W, et al. A novel cytoprotective peptide protects mesenchymal stem cells against mitochondrial dysfunction and apoptosis induced by starvation via Nrf2/Sirt3/FoxO3a pathway. Journal of Translational Medicine. 2017;**15**(1):33

[72] Zeng Y, Zheng L, Yang Z, Yang C, Zhang Y, Li J, et al. Protective effects of cyclic helix B peptide on aristolochic acid induced acute kidney injury. Biomedicine & Pharmacotherapy. 2017;**94**:1167-1175

A Transmembrane Single-Polypeptide-Chain (sc) Linker to Connect the Two G-Protein–Coupled Receptors in Tandem and the Design for an *In Vivo* Analysis of Their Allosteric Receptor-Receptor Interactions

Toshio Kamiya, Takashi Masuko,
Dasiel Oscar Borroto-Escuela, Haruo Okado and
Hiroyasu Nakata

Additional information is available at the end of the chapter

http://dx.doi.org/10.5772/intechopen.71930

Abstract

A transmembrane (TM) single-polypeptide-chain (sc) linker can connect two G-protein–coupled receptors (GPCRs) in tandem. The priority of a gene-fusion strategy for any two class A GPCRs has been demonstrated. In the striatal function, dopamine (DA) plays a critical role. In the striatum, how the GPCR for adenosine, subtype A_{2A} ($A_{2A}R$), contributes to the DA neurotransmission in the "volume transmission"/dual-transmission model has been studied extensively. In addition to the fusion receptor, i.e., the prototype $scA_{2A}R/D_2R$ complex (the GPCR for DA, subtype D_2), several types were created and tested experimentally. To further elucidate this *in vivo*, we designed a new molecular tool, namely, the supermolecule $scA_{2A}R/D_2R$. Here, no experiments on its expression were done. However, the TM linker to connect the nonobligate dimer as the transient class A GPCR nanocluster that has not been identified at the cell surface membrane deserves discussion through $scA_{2A}R/D_2R$. Supramolecular designs, are experimentally testable and will be used to confirm *in vivo* the functions of the two GPCRs interactive in such a low specific signal to the nonspecific noise (S/N) ratio in the neurotransmission in the brain. The sc also has, at last, become straightforward in the field of GPCRs, similar to in the field of antibody.

Keywords: oligomerization, adenosine A_{2A} receptor, dopamine D_2 receptor, receptor allostery, fusion protein, striatum, supramolecular protein assembly

1. Introduction

Dopamine (DA) [1] plays a critical part in the function in the striatum of the basal ganglia [2, 3]. In striatal DA neurotransmission, how the G-protein–coupled receptor (GPCR) for adenosine, subtype A_{2A} ($A_{2A}R$) [4, 5], works in the "volume transmission"/dual-transmission model [6] was explored previously [7, 8]. Moreover, the prototype single-polypeptide-chain (sc) heterodimeric $A_{2A}R/D_2R$ complex (the GPCR for DA, subtype D_2 [9]) [10] (**Figures 1–3**), a fusion receptor, and several other types were created and tested experimentally [11]. Supermolecules were also designed, none of which were constructed or tested [12], while referring to a relationship between nanoscale surface curvature and surface-bound protein oligomerization [13–15]. Here, the transmembrane (TM) linker to connect the nonobligate dimer will be discussed based on $scA_{2A}R/D_2R$.

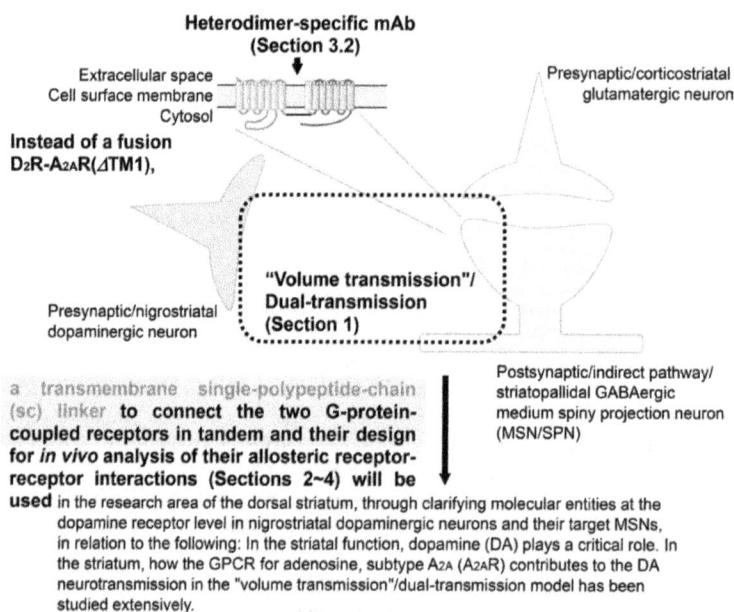

Figure 1. A transmembrane single-polypeptide-chain (sc) linker to connect two G-protein–coupled receptors in tandem. See text for details.

2. Making the single-polypeptide-chain to compensate the weak affinity of the two molecules

2.1. Glycine-glycine-glycine-glycine-serine (G_4S) linker

Approximately 30 years ago, a 15–amino acid linker [glycine-glycine-glycine-glycine-serine $(G_4S)_3$] was adopted to form a variable region fragment (Fv) analogue connected as a

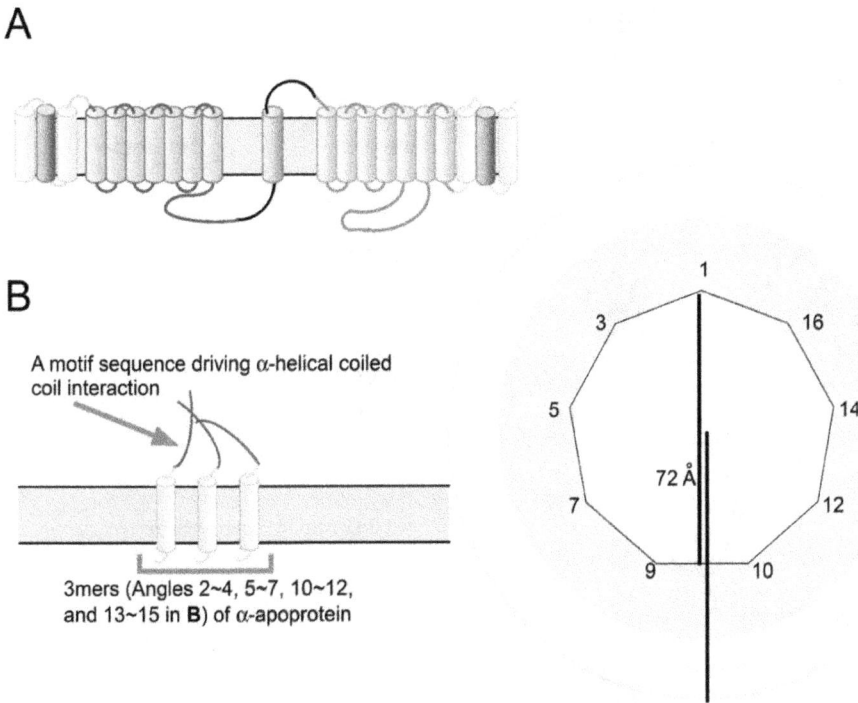

Figure 2. A supermolecule of an 'exclusive' dimeric GPCR with the oil-fence–like structure. (A) Using the light-harvesting antenna complex from *Thermochromatium tepidum*, the C-ter of an α-apoprotein (Angle 1 in B, in dotted line in gray) of two (Angles 1 and 16), each of which is concatenated in tandem through a CD4 transmembrane (TM) region in purple (in darker gray for a printed version in black and white) between Angles 1 and 16; CD4 is well known to make no formation of dimer itself, fused to the N-ter of TM helix 1 of the human prototype $scA_{2A}R/D_{2L}R$, i.e., $A_{2A}R$-*odr4*TM-$D_{2L}R$ colored (in the same gray as in Figure 1 for a printed version in black and white), and its C-ter of TM helix 7 fused to the N-ter of another α-apoprotein (Angle 8) of other two (Angles 8 and 9 here in B, as given angles of the complex, in dotted line in gray), i.e., the α-apoprotein-CD4TM-α-apoprotein-scGPCR-α-apoprotein-CD4TM-α-apoprotein fusion, is shown. Its expression as a fusion with the remaining ~12 α-apoproteins (presumed to be 16 mer originally in total), i.e., four 3mers (Angles 2–4, 5–7, 10–12, and 13–15 in B) of α-apoprotein fused to a motif sequence driving α-helical coiled coil interaction as lines in blue (in darker gray for a printed version in black and white) on the left in B because four 4mers and eight 2mers cannot exclude vacant supermolecules, could form a unified complex, with ~16 wild-type β-apoproteins (plus the translation initiation methionine) in total and pigments, thus surrounding $scA_{2A}R/D_{2L}R$. (B) The light-harvesting antenna complex (LH1, a gray circle) from *Thermochromatium tepidum* consists of 16 mer of the α-apoprotein [61 amino acids, with the intracellular N-terminus (N-ter); for clarity, instead of a hexadecagon (not shown here) with each angle numbered, some are shown as a nonagon] packed side by side to form a hollow cylinder of diameter 73–82 Å and the 16 helical β-apoproteins (46 amino acids plus the translation initiation methionine, with the intracellular N-ter) of an outer cylinder of diameter 96–105 Å, together with light-absorbing pigments (not shown here) 32 bacteriochlorophyll *a* (Bchl *a*) and 16 carotenoids (spirilloxanthin, Spx) [77]. Lines of 72 Å are shown in black. The figure art (A) is shown and was drawn and adapted from the published figure, i.e., the figure art (lower left) for $A_{2A}R$-*odr4*TM-$D_{2L}R$ in Fig. 1A (pp. 140), in our previous report [10], with written permission of the copyright owner, the Japan Society for Cell Biology. Only small but biologically important modifications were introduced.

single polypeptide chain (single-chain Fv), consisting of the heavy- and light-chain variable regions (V_H and V_L) of a monoclonal antibody (mAb). This was successfully produced in *Escherichia coli* or bacteriophage by protein engineering based on the crystallographic analysis of the antigen-binding fragments (Fabs) of antibodies, i.e., the carboxy-terminus

Graphical Abstract

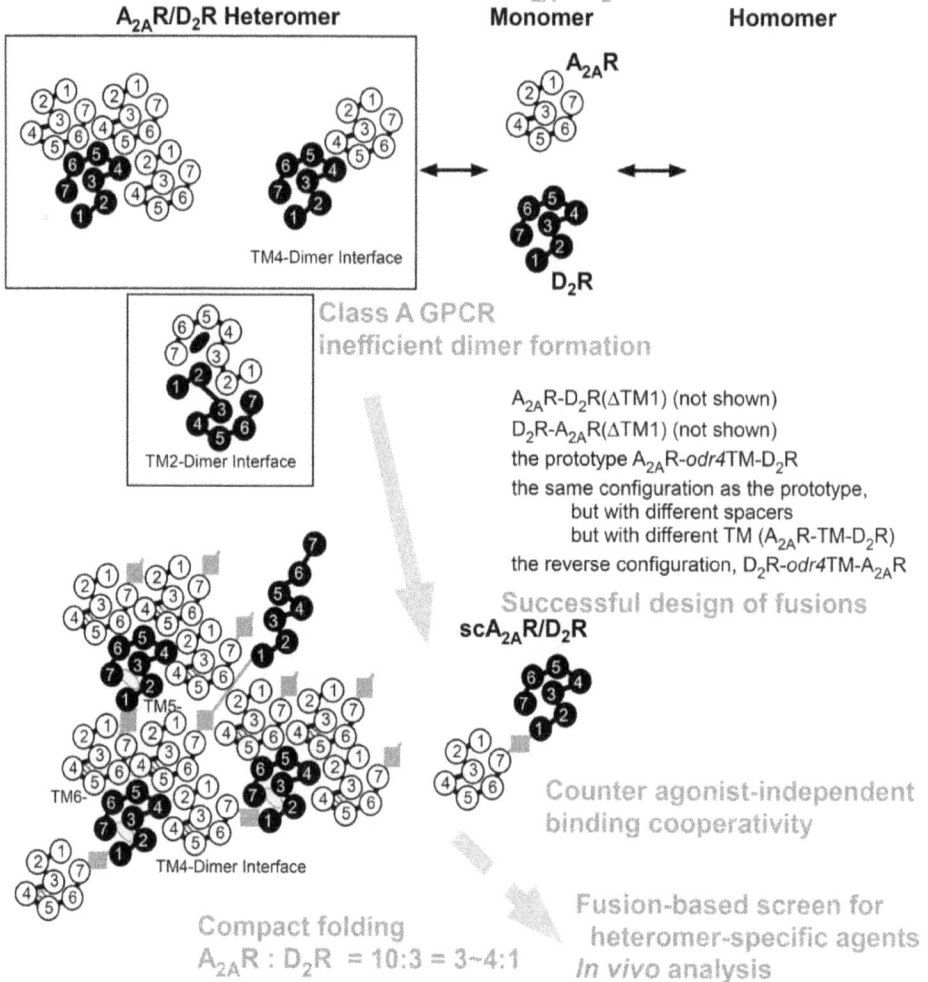

Figure 3. Transient class A GPCR nanocluster. TM helices (transverse sections) are shown as circles [numbered, in black (D_2R) or white ($A_{2A}R$)]. The $A_{2A}R$ and D_2R ligands are also shown as black and light gray ovals, respectively. *odr4*TM in the sc$A_{2A}R/D_2R$ is shown in a gray rectangle. This figure is from the figure art of the graphical abstract in our previous report [11], with written permission of the copyright owner, Elsevier Inc.

(C-ter) of the V_H domain and the amino-terminus (N-ter) of the V_L domain, being at a distance of ≈3.5 nm [16, 17]. The use of a spacer (linker sequence) of ~30 amino acids, including G_4S, between the two proteins was established after that, giving $(G_4S)_3$: 3.5 nm [18]. As a compensation for the weak interactions between the two proteins, a tandem is especially useful. Under the molecular dynamics and the organizing principles of the plasma membrane [19], depending on the GPCR monomer-dimer dynamic equilibrium that is characterized by single-molecule imaging to date [20], the endocytosis of the GPCR is mediated by the clathrin-coated pit machinery [21]. The clathrin coats on the endosome vesicles resemble the architecture of a soccer ball, and each clathrin that forms a three-legged structure assembles into typical polyhedral cages, with an inscribed or circumscribed circle width diameter of approximately 25 nm ([11]: graphical abstract) (**Figure 3**). At issue are the tuning and amplitude of GPCR oligomerization. Using a $\beta 2V2R$ receptor chimera [the class A GPCR β_2 adrenergic receptor where the C-ter tail was exchanged for the class B GPCR vasopressin type 2 receptor (V2R) C-ter] [22], the existence, functionality, and architecture of internalized class B GPCR complexes, called supercomplexes or "megaplexes," resulting in sustained signaling, are reported to consist of a single GPCR, β-arrestin, and G protein. Additionally, GPCR-mediated extracellular signal-regulated kinase (ERK) activation is classified into two modes, including an early, β-arrestin–independent one, which may correspond to nanocluster activation at the cell surface membrane, and a late, β-arrestin–dependent one, which prolongs ERK activity. Nanoclusters are transient dynamic structures that are assembled by the lipid-anchored proteins [23]. With regard to Ras proteins, they are arrayed in nanoclusters comprising 6–8 proteins in domains that are 12–22 nm in diameter. However, the transient class A GPCR nanocluster has not been identified at the cell surface membrane.

2.2. Glycine-threonine (GT) linker

In a report by Twomey et al., in order to form the complex between the α-amino-3-hydroxy-5-methyl-4-isoxazolepropionic acid-subtype ionotropic glutamate receptor GluA2 and the auxiliary protein stargazin (STZ), they used "a tandem construct, GluA2-STZ, where the N-ter of STZ was fused to the C-ter of GluA2 by a glycine-threonine (GT) linker."{[24]—(pp. 83, the right column—the second paragraph—line 1)}.

2.3. Glycine-serine-anchored 6– or 30–amino acid (GSxxGS) linker

In a report by Elegheert et al., they used the fusion protein of the $Cbln1_{C1q}$-$GluD2_{ATD}$. Cbln1 is a soluble synaptic organizer molecule with a compact jelly-roll β-sandwich fold and is a member of complement C1q-tumor necrosis factor superfamily that directly binds the ionotropic glutamate receptor $\delta 2$ extracellular N-ter domain GluD2 ATD. Cbln1 also interacts with presynaptic membrane-tethered neurexins that, with postsynaptic neuroligins, make up the transsynaptic bridges spanning the synaptic cleft [25]. Because of the weak affinity of the $Cbln1_{C1q}$-$GluD2_{ATD}$ interaction, they designed a construct that "linked a $Cbln1_{C1q}$ trimer with $GluD2_{ATD}$ into one continuous polypeptide chain by using short six-amino acid linkers (GSELGS and GSASGS in single-letter amino acid code, respectively)" and "by a 30-residue flexible Gly-Gly-Ser $((G_2S)_{10})$ spacer, which may reach ~110 Å in length in a fully extended conformation and would allow quasi-unrestricted conformational sampling of the $GluD2_{ATD}$ by $Cbln1_{C1q}$-fused during the crystallization process." ([25]—supplementary text).

2.4. Transmembrane linker

Levitz et al. took advantage of a gene-fusion strategy used previously for microbial opsins [26] to construct a tandem dimer, where a TM linker connects the C-ter of the first copy of a class C GPCR, the metabotropic glutamate receptor mGluR2, to the N-ter of a second [27]. The gene-fusion strategy of Kleinlogel et al. proved to be useful in optogenetics and for any two class A GPCR rhodopsins, using "the light-activated microbial rhodopsins": Channelrhodopsin-2 (ChR2) derived from *Chlamydomonas reinhardtii* ["a cation-permeable channel that enables cell depolarization (neuronal activation) in response to blue light"] and halorhodopsin derived from *Natromonas pharaonis* (NphR or Halo) ["a chloride pump that enables cell hyperpolarization (neuronal silencing) in response to orange light"] [26], i.e., the intracellular C-ter of ChR2 through β helix (the 105-amino-acid N-ter fragment of the β subunit of the rat gastric H^+,K^+-ATPase) fused to the extracellular N-ter of NphR for the "precise co-localization and stoichiometric expression of two different light-gated membrane proteins." However, they found that in the fusion of ChR2(1-309) with βbR (a variant of the inhibitory proton pump bacteriorhodopsin from *Halobacterium salinarum*, containing an additional N-ter TM β helix), the insertion of an enhanced yellow fluorescent protein (EYFP) between the two proteins resulted in its functional expression (ChR2-EYFP-βbR). However, they did not try another β helix itself, the effects of their length, or the reverse type (such as bR-EYFP-βChR2), unlike the TM α helix linkers in our previous report [11]. Interestingly, rational *de novo* computational protein design of the α-helical domain is also reported [28–30].

The points raised in our study about how the $A_{2A}R$ contributes to the DA neurotransmission are addressed in a straightforward manner (**Figure 1**) (Section 3.1). Thus, making the single-polypeptide-chain has also, at last, become straightforward and is no longer an unusual approach to stimulate the weak affinity of two molecules in the field of neuroscience/GPCR, similar to in the field of immunology/monoclonal antibody. Additionally, "the glycan wedge" approach by "a 10-residue glycosylated linker" (ELSNGTDGAS in single-letter amino acid code) arranged between the ATD and ligand-binding domain (LBD) layers "in order to space them apart and disrupt potential mechanical ATD-LBD coupling" [25] is shown to function in inhibiting the association between protomers of the γ-aminobutyric acid (GABA) type B receptor $(GABA_B)[GABA_{B1} (GB1)]/[GABA_{B2} (GB2)]$ heterodimer (via coiled-coil interaction of the cytoplasmic C-ter) but not that of the tetramer (GB2/GB1)-(GB1/GB2) [31].

Furthermore, the DA neurotransmission in the "volume transmission"/dual-transmission model could not physically adopt the supramolecular architectures, such as "the prototypical molecular bridge linking" postsynaptic GluD2 and the presynaptic neurexin, via Cbln1, besides a possible link between the glial $A_{2A}Rs$ and presynaptic or postsynaptic D_2R.

2.5. An enzyme-dependent covalent biotinylation occurs within 10–50 nm of the bait protein

BioID is an affinity purification approach where an *E. coli* BirA biotin-protein ligase, BirAR118G (BirA, with Gly replacing Arg 118), is fused to a bait protein expressed in cells and allows for the isolation and analysis of proximal proteins by streptavidin-based affinity purification and mass spectrometry [32–36]. "Biotinylation by BioID is a mark of proximity and not evidence for physical interaction," and causing "the practical labeling radius of BioID *in vivo* to be ~10 nm" [32]. Thus, similar to the proximity ligation assay (PLA), BioID is a "proximity assay that may detect adjacent proteins that are not true interactor" [36]. Across the cell surface membrane, BirA-dependent covalent biotinylation cannot occur even within 10–50 nm of the bait protein.

3. The transmembrane-linked connection of the nonobligate dimer: $scA_{2A}R/D_2R$

3.1. An approach toward a class A GPCR dimer that is not fully formed

With a model of receptor-receptor interaction to regulate DAergic activity, a functional antagonistic interaction between $A_{2A}R$ and D_2R has been explicated [37–40]. Although allostery in a GPCR heterodimer is demonstrated [41–45], these class A GPCR dimers, unlike other class GPCRs that are fully formed [46], depend on the equilibrium between monomers and dimers [47]. This does not mean that such insufficient class A GPCR dimers or oligomers cannot function *in vivo*. Some types of protein-protein interactions, transient or weaker, "will be found to play an even more important role" in the cells [48, 49].

Interestingly, in the process of rational design and screening, we found that fusion of the two receptors stimulates the receptor dimer formation [10, 11]. In these studies, by fusing the cytoplasmic C-ter of the human brain–type $A_{2A}R$ (that is derived from Dr. Shine's cDNA [50]) ([12]: Fig. S1) through the TM domain of a type II TM protein (Section 4.3) with the extracellular N-ter of D_2R in tandem, we made successful designs for a fusion receptor, single-polypeptide-chain (sc) heterodimeric GPCR complex $A_{2A}R/D_2R$ [10, 11]. However, the resulting prototype $scA_{2A}R/D_{2L}R$ ($D_{2L}R$, the long form of D_2R) has a compact folding, i.e., a fixed stoichiometry (the apparent ratios of $A_{2A}R$ to D_2R binding sites), $A_{2A}R:D_2R = 10:3 = 3–4:1$ ([11]: graphical abstract) (**Figure 3**), and the $scA_{2A}R/D_2R$ expression system shows that the various designed types of functional $A_{2A}R/D_2R$ exist even in living cells, but there is no apparent allostery as a whole. Thus, to further clarify the heteromerization through $scA_{2A}R/D_{2L}R$, we tried to design other fusion proteins so as not to be formed/expressed as higher-order-oligomers, and we called these 'exclusive' monomers or dimers. First, we noted that GPCRs have general features of a TM helix 3 as the structural [at a tilt-angle of 35° to a perpendicular (vertical) line to the cell surface membrane plane]/functional hub and a TM helix 6 moving along 14 Å after activation [51] and of $A_{2A}R$, with a bundle width diameter of approximately 3.6 nm [12]. Thus, using a partner to increase spacing [25] without identifying and specifically blocking the interacting portions between the receptors, we created the designs for nonoligomerized 'exclusive' monomeric $A_{2A}R$ and/or D_2R in order to exclude their dimer/oligomer formation [12] (Sections 4.1.1 and 4.1.2), and did those for the 'exclusive' dimer [12] [Fig. 2, (Section 4.2)]. Such a self-assembled molecular architecture will entirely hold either a monomeric receptor or single dimeric $A_{2A}R/D_2R$ alone, but none of the oligomers. Although we constructed or tested none of these new fusions, we aimed to obtain heterodimer-specific agents using the fusion receptor $scA_{2A}R/D_2R$. Indeed, we can take an example of a universal influenza vaccine that was engineered by fusing two polypeptides. The polypeptides originally resulted from the limited proteolysis of the native conformation of the hemagglutinin (HA), a trimeric membrane protein of influenza viruses, and their conformations were retained and stabilized in the vaccine engineering [52]. Such 'exclusive' forms of single-chain dimers are useful analytical tools, allowing us to address this point. Additionally, in the postsynaptic striatopallidal γ-aminobutyric acid (GABA)-ergic medium spiny projection neurons (the indirect pathway), expressing both $A_{2A}R$ and D_2R, an *in vivo* analysis of knock-in mice of the $scA_{2A}R/D_2R$ would elucidate the functional $A_{2A}R/D_2R$ with the antagonism [12], thus, to confirm us, *in vivo*, that such a low S/N ratio interaction between $A_{2A}R$ and $D_{2L}R$ functions in the DA neurotransmission in the striatum.

The heterodimeric interaction of $A_{2A}R$ with D_2R depends, in part, on the cytoplasmic C-ter region of $A_{2A}R$ [53]. Although the three-dimensional (3D) structures of the GPCR heterodimer also remains unresolved crystallographically [39], a possible monovalent agent acting on GPCR heterodimers was made reference to, and a screen needs developing to this end [54]. Our goal is to prove that the above-mentioned scGPCR-based screen is such a system. Here, we aim to obtain such heterodimer-specific agents [12], using a supramolecularly [55–57] designed fusion receptor, $scA_{2A}R/D_2R$, i.e., nonoligomerized 'exclusive' monomer (Section 4.1) and dimer (Section 4.2) of the receptors, and the above-mentioned *in vivo* analysis of the functional antagonistic $A_{2A}R/D_2R$. The possible occurrence of an unsuitable folding into a 3D structure, such that the resulting receptor exhibits lower or false activity, should be avoided. This is attributed to the interaction between a single 'exclusive' form of either the receptor monomer or $scA_{2A}R/D_2R$ and the surrounding fence-like architecture, while considering a single bond between two carbon atoms, with a C—C covalent bond with a distance of 1.5 Å (Section 4.3).

3.2. The molecular populations of the $A_{2A}R/D_{2L}R$ species in cell membranes

To illustrate this point, for clarity, let us consider epitopes generated only in heterodimeric $A_{2A}R/D_{2L}R$, but not in monomeric (and/or homodimeric) $A_{2A}R$ or $D_{2L}R$ [11] (**Figure 1**), which is in accordance with findings on the existence of agonistic/antagonistic (active/inactive-state-specific) or dimer-specific antibodies (nanobody) ([12]: Table 1). A broad and extremely potent human immunodeficiency virus (HIV)-specific mAb, termed 35O22, is reported, which binds the gp41–gp120 interface of the viral envelope glycoprotein (trimer of gp41–gp120 heterodimers) [58] ([12]: Table 1). This dimer-specific mAb was obtained despite not being immunized. The existence of virus-neutralizing mAbs, such as 35O22, which recognizes HIV-1 gp41–gp120 interface [58], and 2D22, locking the dimeric envelope proteins of dengue virus type 2 [59], is suggestive of that of the heterodimer-specific mAb that we are interested in. Additionally, transient nucleotide-bound β2-Gs species that are distinct from known structures are revealed [60]. Thus, the expression of homogeneous molecular species, either monomer or dimer, but not the mixture of both, followed by their membrane preparation appropriate to our needs, is necessary and worthy to be addressed experimentally.

4. Architecture of a transmembrane-linked scA$_{2A}$R/D$_2$R

4.1. Supramolecular monomer

4.1.1. Protein assembly regulation and 'exclusive' monomers, supramolecularly designed using the Cε2 domain of IgE-Fc: the scA$_{2A}$R/D$_2$R-transmembrane linker makes both receptors stay away from each other

The molecular entity of the allosteric modulation of A$_{2A}$R/D$_2$R remains unresolved. To solve the insufficiency of the dimer formation of A$_{2A}$R/D$_{2L}$R, various scA$_{2A}$R/D$_2$R constructs, with spacers between the two receptors, were created (**Figure 3**). Successful designs of fusions, A$_{2A}$R-D$_2$R(ΔTM1) (not shown), D$_2$R-A$_{2A}$R(ΔTM1) (**Figure 1**), the prototype A$_{2A}$R-*odr*4TM-D$_2$R, fusions, which have the same configuration as the prototype, but with different spacers, and the same configuration as the prototype, but with different TM (A$_{2A}$R-TM-D$_2$R), and the reverse configuration, D$_2$R-*odr*4TM-A$_{2A}$R, were designed. Using whole cell binding assays, the constructs were examined for their binding activity. Two papers [61, 62] inspired us to also design the following fusions ([12]: Fig. 1B): first, the conversion between a single- and two-antigen binding brought by using a hinge/domain in the designed antibody was reported. Then, in models of viral fusogenic proteins, both steric hindrance and conformational changes, i.e., negative cooperativity, were referred to. Accordingly, we took advantage of the structure of the complete Fc fragment (Fc) of immunoglobulin (Ig) E, including the Cε2 domains, which is a compact, bent conformation ([63]: pp. 205, the right column-line 2 from the bottom; [64]) (the human IgE has a rigid Cε2 domain of Fc portion, instead of lacking a flexible hinge region, in contrast to other class/subclasses, such as IgG1. Upon binding of an allergen to the IgE that is already bound to the high-affinity receptor FcεRI, the antigen binding fragment (Fab) portion transduces it to the FcεRI [64]). Thus, the expression of the 'exclusive' monomeric GPCRs linked with the transmembrane plus human Cε2 domain (here in a loop), i.e., the C-ter of the *odr*4TM of the prototype scA$_{2A}$R/D$_{2L}$R fused to the N-ter of Cε2 and its C-ter fused to the N-ter of the D$_{2L}$R, could separate from each other. Whereas the prototype A$_{2A}$R-*odr*4TM-D$_2$R stimulates the dimerization of A$_{2A}$R and D$_2$R, this type of Cε2-intervening scA$_{2A}$R/D$_2$R makes both receptors push out each other, resulting in two 'exclusive' monomers. To this end, additional bulky molecules at both the N-ter and the C-ter of the fusion would be required.

4.1.2. Another 'exclusive' monomeric GPCR with the oil-fence–like structure: a supermolecule using transmembrane apoproteins from a bacterial light-harvesting antenna complex

The human A$_{2A}$R structure PDB 3EML [12] has a bundle width diameter of approximately 36 Å. It was determined by the T4-lysozyme fusion strategy [65], where most of the intracellular loop 3 (Leu209$^{5.70}$–Ala221$^{6.23}$: the Ballesteros-Weinstein numbering scheme is shown in superscript) [66] was replaced with a lysozyme from T4 bacteriophage, and the C-ter tail (Ala317–Ser412) was deleted to improve the likelihood of crystallization. According to recent papers, 'the most complex designed membrane proteins contain porphyrins that catalyze transmembrane electron transfer' [55]. The peripheral light-harvesting antenna complex (LH2) [12] derived from the purple bacterium *Rhodopseudomonas acidophila* (*Rhodoblastus acidophilus*) strain 10050 is made up of both a 9 mer of the transmembrane α-apoprotein (53 amino acids, with the intracellular N-ter) grouped side by side to form a hollow cylinder with a radius of 18 Å and the 9 transmembrane helical β-apoproteins (41 amino acids, with the intracellular N-ter) of an outer cylinder with a radius of 34 Å, together with porphyrin-like, light-absorbing pigments bacteriochlorophyll *a* (Bchl *a*) and carotenoids [67]. Thus, a complex could be formed as an `exclusive' monomeric GPCR with the oil-fence–like structure, by expressing both the C-ter of this α-apoprotein fused to the N-ter of TM helix 1 of GPCR, plus the C-ter of the TM helix 7 fused to the N-ter of another α-apoprotein, and other 7 wild-type α-apoproteins, together with 9 β-apoproteins and pigments [12].

4.2. An `exclusive' dimeric GPCR with the oil-fence–like structure: a supramolecular dimer

Using the LH complex from *Thermochromatium tepidum*, the C-ter of the second copy (**Figure 2**: Angle 1 in B, in dotted line in gray) of two α-apoproteins (Angles 1 and 16), each of which is concatenated in tandem through the human leukocyte antigen (cluster of differentiation, CD) CD4 TM region (in purple between Angles 1 and 16; CD4 is well known to make no formation of a dimer itself), fused to the N-ter of TM helix 1 of the human prototype scA$_{2A}$R/D$_{2L}$R, i.e., A$_{2A}$R-*odr*4TM-D$_{2L}$R (colored), and its C-ter of TM helix 7 fused to the N-ter of another α-apoprotein (Angle 8) of the other two (Angles 8 and 9 herein in B, as given Angles of the complex, in dotted line in gray), i.e., the α-apoprotein-CD4TM-α-apoprotein-scGPCR-α-apoprotein-CD4TM-α-apoprotein fusion, is shown. Its expression as a fusion, with the remaining ~12 α-apoproteins (presumed to be 16 mer originally in total), i.e., four 3mers (Angles 2–4, 5–7, 10–12, and 13–15 in B) of α-apoprotein fused to a motif sequence driving α-helical coiled coil interaction (as blue lines on the left in B) because four 4mers and eight 2mers cannot exclude vacant supermolecules, could form a unified complex, with ~16 wild-type β-apoproteins (plus the translation initiation methionine) in total and pigments, thus surrounding scA$_{2A}$R/D$_{2L}$R.

4.3. Predicting the interaction between the TM linker, *odr*4TM, in prototype scA$_{2A}$R/D$_2$R and TMs of its surrounding (fence-like) LH proteins

Protein-protein interactions can be classified on the basis of their binding affinities [48, 49, 68, 69]: by definition, unlike permanent interactions with high affinities (K_d in the nM range), proteins interacting transiently, either weakly or strongly, show a fast bound-unbound equilibrium, with K_d values typically in the μM range or less. In our previous reports [10, 11], among the GPCR protein-protein interactions, such as the disulfide bond formation of the N-ter, coiled-coil interaction of the cytoplasmic C-ter, and TM interaction [70], a type II TM protein with a cytoplasmic N-ter segment, single TM, and extracellular C-ter tail, i.e., the *Caenorhabditis elegans* accessory protein of odorant receptor (*odr*4) [71], was first selected for a connection between the N-ter receptor half (A$_{2A}$R) and the C-ter receptor half (D$_{2L}$R) of the scA$_{2A}$R/D$_{2L}$R. Then, it was demonstrated that the insertion of some other TM sequence, instead of the *odr*4TM sequence, works similarly or that it does not have to be *odr*4TM to work, using the scA$_{2A}$R/D$_{2L}$R designed with another TM of a type II TM protein, the human low-affinity receptor for IgE designated CD23 (Section 3.1).

Among the Smart blast search of *odr*4TM hits, a hit of a photosynthetic protein [of photosystem (PS) II reaction center (RC)] is impressive to us because of the use of its surrounding apoproteins in the bacterial LH complex in the supramolecularly designed scA$_{2A}$R/D$_2$R, as described above [12]. The Blastp search of the *odr*4TM against all nonredundant GenBank databases and the human genome gave no hits for human CD23.

On the other hand, CD23 is highly conserved among mammals. The Smart blast search suggests that human CD23TM is selected as a preferable experimental design due to little relationship between the two protein TMs.

In photosynthesis, certain protein complexes, such as the PS (or the LH1 and the RC), build the highly efficient, light-induced charge separation across the membrane, followed by electron transport. More strictly, PSI forms supercomplexes with the RC and the light-harvesting proteins, i.e., PSI-type RC and LHCI, and PSII supercomplexes are PSII-type RC, core light-harvesting proteins (CP43 and CP47), and peripheral LHCII in aerobic photosynthesis (algae, cyanobacteria, and plants); PS is a type RC, core LH1, and peripheral LH2 in anaerobic photosynthesis (purple bacteria). Whether this quality of photosynthetic light reactions is negligible or not in our 'exclusive' supramolecular forms has not been tested. The rearrangement of each molecule in this architecture or exchange with similar but unrelated molecules [72] would be necessary to inhibit this completely and surround the prototype scA$_{2A}$R/D$_2$R.

In a recent report on PSII biogenesis, it is shown that, unlike its cyanobacterial counterpart, the PSII RC protein D1 C-terminal processing enzyme of a land plant *Arabidopsis* is essential for assembling functional PSII core complexes, dimers, and PSII supercomplexes [73], demonstrating a discrepancy in PSII protein assembly [72] between cyanobacterium and *Arabidopsis*. Thus, whether the LH is attached and expressed to form our supermolecules in animal cells requires testing. The D1 and D2 subunits of the PSII and the M and L subunits of the bacterial photosynthetic RC are members of the Photo RC (cl08220) protein superfamily. However, these proteins differ in their number of TM helices [12, 74–77]. In addition, for the following three reasons, it is persuasive to predict the interaction between *odr*4TM in prototype scA$_{2A}$R/D$_2$R and TMs of its surrounding (fence-like) LH proteins: (1) based on findings that the RC affects the LH1 complex shape [78], (2) and that it remains unknown what structural features lead to the consequent differences in the nonameric and octameric apoprotein assemblies in LH2, respectively, from *Rhodopseudomonas acidophila* and another purple bacterium, *Rhodospirillum (Phaeospirillum) molischianum* [79], thus meaning that it remains unknown why LH1 surrounds the RC but LH2 does not, (3) and furthermore, due to the difficulty in membrane-protein topology prediction itself, up to 25% precision, "to predict interaction sites from sequence information only" {[80]: supporting information—[Table 9. List of PDB IDs of all monomers (Test-set 1)] (pp. 38, line 27), human A$_{2A}$R (PDB: 3EML) ([12]: Fig. 2A) (Section 3.1.2) defined as a monomer}.

Thus, as a source of information for the prediction of the protein-protein interactions [81], multiple sequence alignments were analyzed between the *C. elegans odr4*TM or human CD23 TM and TMs of core RC proteins (L/M/H) of *T. tepidum* or *Rhodopseudomonas* (Blastochloris) *viridis* [12]. It indicates no relationship [12], suggesting that the $\alpha\beta$ subunits for the LH surrounding scA$_{2A}$R/D$_2$R do not affect their core GPCR itself substituted for the core RC, even if each protein in this supermolecule is assembled to form it. Whether alternative interactions between core GPCRs and surrounding LH subunits in our supermolecules exist remains unknown because our designs are only back-of-envelope sketches and also, 3D-models, especially of membrane proteins, have their limitations [12, 82–84].

5. Summary and outlook

The supramolecular designs of transmembrane-linked scA$_{2A}$R/D$_{2L}$R, 'exclusive' monomers and dimers using the Cϵ2 domain of IgE-Fc or apoproteins of the bacterial light-harvesting antenna complex, allowing us to express the class A GPCR by receptor protein assembly regulation, i.e., the selective monomer/nonobligate dimer formation, are experimentally testable and will be used to confirm, *in vivo*, that such low S/N ratio interaction between A$_{2A}$R and D$_{2L}$R functions in the dopamine neurotransmission in the striatum.

Acknowledgements

We thank Drs. O. Saitoh (Nagahama Institute of Bio-Science and Technology, Nagahama) and K. Yoshioka (Kanazawa University, Kanazawa) for their respective contributions [11], Mr. M. Woolfenden for proofreading the English of a part of this manuscript, and Dr. K. Fuxe (Karolinska Institutet, Stockholm) and Dr. H. Saya (Keio University, Tokyo) for encouragement. This work was supported, in part, by grants for Scientific Research from the Ministry of Education, Culture, Sports, Science and Technology (MEXT) (#14657595, 13670109: to T.K., O.S., and H.N.). This work was also supported, in part, by grants of intramural budget (TMIN to T.K., O.S., and H.N.). T.M. paid the cost of publication fee associated with open access journal. T.M. was also supported by both the "Academic Frontier" Project for Private Universities: matching fund subsidy from MEXT, 2005-2007, and the MEXT-supported Program for the Strategic Research Foundation at Private Universities, 2014-2018 (#S1411037). H.N. was also supported by a CREST program of the Japan Science and Technology Agency.

Conflicts of interest

The authors declare no competing financial interests.

The founding sponsors had no role in the design of the study, in the collection, analyses, or interpretation of data, in the writing of the manuscript, and in the decision to publish the results.

Notes

Kinki University, to which Takashi Masuko has been affiliated, changed its English name to Kindai University in the year 2016 (URL: http://www.kindai.ac.jp/english/about/history.html).

Division of Gene Regulation, Institute for Advanced Medical Research, Keio University School of Medicine, Keio University, Shinjuku-ku, Tokyo, Japan (Toshio Kamiya: Present Address).

Author contributions

T.K. contributed to the planning and interpretation of all of the experiments and conducted all of them and performed the design [at The University of Tokyo General Library (Hongo, Bunkyo-ku, Tokyo)] [from April 1, 2013 to date (this submission)] of vectors for the nono-ligomerized 'exclusive' monomer and dimer, experiments such as antibody screening at TMIN [11], the *in silico* search and wrote the paper; O.S. contributed to the interpretation of all of the experiments done at TMIN [11]; T.M. performed immunization and cell fusion to make hybridomas at Kinki University and, with T.K., wrote the paper; H.O. performed the design (It started from December 2014) of the vectors for the *in vivo* analysis and, with T.K., wrote the paper; K.F. supervised the project of the electrophysiological studies (It started from November 2007) [11]; with T.K., D.O.B.-E. analyzed the A_{2A}R SNP and wrote the paper; and H.N. supervised the overall project by the end of March 2007 and also contributed to the interpretation of all of the experiments by the end of March 2007. All the authors commented on the manuscript, except for H.N., who is, at present, under a departure from research activity.

Author details

Toshio Kamiya[1,2,3*], Takashi Masuko[3], Dasiel Oscar Borroto-Escuela[4], Haruo Okado[5] and Hiroyasu Nakata[1]

*Address all correspondence to: kamiya@z2.keio.jp

1 Department of Molecular Cell Signaling, Tokyo Metropolitan Institute for Neuroscience, Fuchu, Tokyo, Japan

2 Department of Neurology, Tokyo Metropolitan Institute for Neuroscience, Fuchu, Tokyo, Japan

3 Cell Biology Laboratory, School of Pharmaceutical Sciences, Kinki University, Higashi-Osaka, Osaka, Japan

4 Department of Neuroscience, Karolinska Institutet, Stockholm, Sweden

5 Neural Development Project, Tokyo Metropolitan Institute of Medical Science, Setagaya-ku, Tokyo, Japan

References

[1] Spano PF, Trabucchi M, Di Chiara G. Localization of nigral dopamine-sensitive adenylate cyclase on neurons originating from the corpus striatum. Science. 1977;**196**:1343-1344. DOI: 10.1126/science.17159

[2] Kreitzer AC, Malenka RC. Striatal plasticity and basal ganglia circuit function. Neuron. 2008;**60**:543-554. DOI: 10.1016/j.neuron.2008.11.005

[3] Donahue CH, Kreitzer AC. A direct path to action initiation. Neuron. 2015;**88**:240-241. DOI: 10.1016/j.neuron.2015.10.013

[4] Glaser T, Cappellari AR, Pillat MM, Iser IC, Wink MR, Battastini AM, Ulrich H. Perspectives of purinergic signaling in stem cell differentiation and tissue regeneration. Purinergic Signal. 2012;**8**:523-537. DOI: 10.1007/s11302-011-9282-3

[5] Ralevic V, Burnstock G. Receptors for purines and pyrimidines. Pharmacological Reviews. 1998;**50**:413-492 http://pharmrev.aspetjournals.org/content/50/3/413

[6] Wall NR, De La Parra M, Callaway EM, Kreitzer AC. Differential innervation of direct- and indirect-pathway striatal projection neurons. Neuron. 2013;**79**:347-360. DOI: 10.1016/j.neuron.2013.05.014

[7] Fuxe K, Cintra A, Agnati LF, Härfstrand A, Goldstein M. Studies on the relationship of tyrosine hydroxylase, dopamine and cyclicAMP-regulated phosphoprotein-32 immuno-reactive neuronal structures and D1 receptor antagonist binding sites in various brain regions of the male rat-mismatches indicate a role of d1 receptors in volume transmission. Neurochemistry International. 1988;**13**:179-197. DOI: 10.1016/0197-0186(88)90054-X

[8] Navarro G, Borroto-Escuela DO, Fuxe K, Franco R. Purinergic signaling in Parkinson's disease. Relevance for treatment. Neuropharmacology. 2016;**104**:161-168. DOI: 10.1016/j.neuropharm.2015.07.024

[9] Missale C, Nash SR, Robinson SW, Jaber M, Caron MG. Dopamine receptors: From structure to function. Physiological Reviews. 1998;**8**:89-225 http://physrev.physiology.org/content/78/1/189.long

[10] Kamiya T, Saitoh O, Nakata H. Functional expression of single-chain heterodimeric G-protein-coupled receptor for adenosine and dopamine. Cell Structure and Function. 2005;**29**:139-145. DOI: 10.1247/csf.29.139

[11] Kamiya T, Yoshioka K, Nakata H. Analysis of various types of single-polypeptide-chain (sc) heterodimeric $A_{2A}R/D_2R$ complexes and their allosteric receptor-receptor interactions. Biochemical and Biophysical Research Communications. 2015;**456**:573-579. DOI: 10.1016/j.bbrc.2014.11.098

[12] Kamiya T, Masuko T, Borroto-Escuela DO, Okado H, Nakata H. Design: An assay based on single-polypeptide-chain heterodimeric $A_{2A}R/D_2R$ and non-oligomerized fusions for *in vivo* analysis of their allosteric receptor-receptor interactions. bioRxiv. 2016. DOI: 10.1101/065250

[13] Chernomordik LV, Kozlov MM. Mechanics of membrane fusion. Nature Structural & Molecular Biology. 2008;**15**:675-683. DOI: 10.1038/nsmb.1455

[14] Kurylowicz M, Paulin H, Mogyoros J, Giuliani M, Dutcher JR. The effect of nanoscale surface curvature on the oligomerization of surface-bound proteins. Journal of the Royal Society Interface. 2014;**11**:20130818. DOI: 10.1098/rsif.2013.0818

[15] Soubias O, Teague WE Jr, Hines KG, Gawrisch K. The role of membrane curvature elastic stress for function of rhodopsin-like G protein-coupled receptors. Biochimie. 2014;**107**:28-32. DOI: 10.1016/j.biochi.2014.10.011

[16] Huston JS, Levinson D, Mudgett-Hunter M, Tai M-S, Novotný J, Margolies MN, Ridge RJ, Bruccoleri RE, Haber E, Crea R, Oppermann H. Protein engineering of antibody binding sites: Recovery of specific activity in an anti-digoxin single-chain Fv analogue produced in *Escherichia coli*. Proceedings of the National Academy of Sciences of the United States of America. 1988;**85**:5879-5883 http://www.pnas.org/content/85/16/5879.long

[17] McCafferty J, Griffiths AD, Winter G, Chiswell DJ. Phage antibodies: Filamentous phage displaying antibody variable domains. Nature. 1990;**348**:552-554. DOI: 10.1038/348552a0

[18] Sambrook J, Russell DW. Molecular Cloning: A Laboratory Manual. 3rd ed. New York: Cold Spring Harbor Laboratory Press; 2001 p. 18.120, 18.122

[19] Kusumi A, Fujiwara TK, Chadda R, Xie M, Tsunoyama TA, Kalay Z, Kasai RS, Suzuki KGN. Dynamic organizing principles of the plasma membrane that regulate signal transduction: Commemorating the fortieth anniversary of Singer and Nicolson's fluid-mosaic model. Annual Review of Cell and Developmental Biology. 2012;**28**:215-250. DOI: 10.1146/annurev-cellbio-100809-151736

[20] Kasai RS, Suzuki KGN, Prossnitz ER, Koyama-Honda I, Nakada C, Fujiwara TK, Kusumi A. Full characterization of GPCR monomer-dimer dynamic equilibrium by single molecule imaging. The Journal of Cell Biology. 2011;**192**:463-480. DOI: 10.1083/jcb.201009128

[21] Pierce KL, Lefkowitz RJ. Classical and new roles of β-arrestins in the regulation of G-protein-coupled receptors. Nature Reviews Neuroscience. 2001;**2**:727-733. DOI: 10.1038/35094 577

[22] Thomsen AR, Plouffe B, Cahill TJ 3rd, Shukla AK, Tarrasch JT, Dosey AM, Kahsai AW, Strachan RT, Pani B, Mahoney JP, Huang L, Breton B, Heydenreich FM, Sunahara RK, Skiniotis G, Bouvier M, Lefkowitz RJ. GPCR-G protein-β-arrestin super-complex mediates sustained G protein Signaling. Cell. 2016;**166**:907-919. DOI: 10.1016/j.cell.2016.07.004

[23] Nussinov R, Jang H, Tsai CJ. Oligomerization and nanocluster organization render specificity. Biological Reviews of the Cambridge Philosophical Society. 2015;**90**:587-598. DOI: 10.1111/brv.12124

[24] Twomey EC, Yelshanskaya MV, Grassucci RA, Frank J, Sobolevsky AI. Elucidation of AMPA receptor-stargazin complexes by cryo-electron microscopy. Science. 2016;**353**:83-86. DOI: 10.1126/science.aaf8411

[25] Elegheert J, Kakegawa W, Clay JE, Shanks NF, Behiels E, Matsuda K, Kohda K, Miura E, Rossmann M, Mitakidis N, Motohashi J, Chang VT, Siebold C, Greger IH, Nakagawa T, Yuzaki M, Aricescu AR. Structural basis for integration of GluD receptors within synaptic organizer complexes. Science. 2016;353:295-299. DOI: 10.1126/science.aae0104

[26] Kleinlogel S, Terpitz U, Legrum B, Gökbuget D, Boyden ES, Bamann C, Wood PG, Bamberg E. A gene-fusion strategy for stoichiometric and co-localized expression of light-gated membrane proteins. Nature Methods. 2011;8:1083-1088. DOI: 10.1038/nmeth.1766

[27] Levitz J, Habrian C, Bharill S, Fu Z, Vafabakhsh R, Isacoff EY. Mechanism of assembly and cooperativity of homomeric and heteromeric metabotropic glutamate receptors. Neuron. 2016;92:143-159. DOI: 10.1016/j.neuron.2016.08.036

[28] Brunette TJ, Parmeggiani F, Huang P-S, Bhabha G, Ekiert DC, Tsutakawa SE, Hura GL, Tainer JA, Baker D. Exploring the repeat protein universe through computational protein design. Nature. 2015;528:580-584. DOI: 10.1038/nature16162

[29] Doyle L, Hallinan J, Bolduc J, Parmeggiani F, Baker D, Stoddard BL, Bradley P. Rational design of α-helical tandem repeat proteins with closed architectures. Nature. 2015;528:585-588. DOI: 10.1038/nature16191

[30] Mondal S, Adler-Abramovich L, Lampel A, Bram Y, Lipstman S, Gazit E. Formation of functional super-helical assemblies by constrained single heptad repeat. Nature Communications. 2015;6:8615. DOI: 10.1038/ncomms9615

[31] Comps-Agrar L, Kniazeff J, Nørskov-Lauritsen L, Maurel D, Gassmann M, Gregor N, Prézeau L, Bettler B, Durroux T, Trinquet E, Pin J-P. The oligomeric state sets GABA$_B$ receptor signalling efficacy. The EMBO Journal. 2011;30:2336-2349. DOI: 10.1038/emboj.2011.143

[32] Kim DI, Birendra KC, Zhu W, Motamedchaboki K, Doye V, Roux KJ. Probing nuclear pore complex architecture with proximity-dependent biotinylation. Proceedings of the National Academy of Sciences of the United States of America. 2014;111:E2453-E2461. DOI: 10.1073/pnas.1406459111

[33] Uezu A, Kanak DJ, Bradshaw TWA, Soderblom EJ, Catavero CM, Burette AC, Weinberg RJ, Soderling SH. Identification of an elaborate complex mediating postsynaptic inhibition. Science. 2016;353:1123-1129. DOI: 10.1126/science.aag0821

[34] Roux KJ, Kim DI, Raida M, Burke B. A promiscuous biotin ligase fusion protein identifies proximal and interacting proteins in mammalian cells. The Journal of Cell Biology. 2012;196:801-810. DOI: 10.1083/jcb.201112098

[35] van Steensel B, Henikoff S. Identification of in vivo DNA targets of chromatin proteins using tethered Dam methyltransferase. Nature Biotechnology. 2000;18:424-428. DOI: 10.1038/74487

[36] Yao Z, Petschnigg J, Ketteler R, Stagljar I. Application guide for omics approaches to cell signaling. Nature Chemical Biology. 2015;11:387-397. DOI: 10.1038/nchembio.1809

[37] Ferre S, von Euler G, Johansson B, Fredholm BB, Fuxe K. Stimulation of high-affinity adenosine A₂ receptors decreases the affinity of dopamine D₂ receptors in rat striatal membranes. Proceedings of the National Academy of Sciences of the United States of America. 1991;**88**:7238-7241 http://www.pnas.org/content/88/16/7238.long

[38] Agnati LF, Ferré S, Lluis C, Franco R, Fuxe K. Molecular mechanisms and therapeutical implications of intramembrane receptor/receptor interactions among heptahelical receptors with examples from the striatopallidal GABA neurons. Pharmacological Reviews. 2003;**55**:509-550. DOI: 10.1124/pr.55.3.2

[39] Cordomi A, Navarro G, Aymerich MS, Franco R. Structures for G-protein-coupled receptor tetramers in complex with G proteins. Trends in Biochemical Sciences. 2015;**40**:548-551. DOI: 10.1016/j.tibs.2015.07.007

[40] Bonaventura J, Navarro G, Casadó-Anguera V, Azdad K, Rea W, Moreno E, Brugarolas M, Mallol J, Canela EI, Lluís C, Cortés A, Volkow ND, Schiffmann SN, Ferré S, Casadó V. Allosteric interactions between agonists and antagonists within the adenosine A₂ₐ receptor-dopamine D₂ receptor heterotetramer. Proceedings of the National Academy of Sciences of the United States of America. 2015;**112**:E3609-E3618. DOI: 10.1073/pnas.1507704112

[41] Wootten D, Christopoulos A, Sexton PM. Emerging paradigms in GPCR allostery: Implications for drug discovery. Nature Reviews Drug Discovery. 2013;**12**:630-644. DOI: 10.1 038/nrd4052

[42] Christopoulos A, Changeux J-P, Catterall WA, Fabbro D, Burris TP, Cidlowski JA, Olsen RW, Peters JA, Neubig RR, Pin J-P, Sexton PM, Kenakin TP, Ehlert FJ, Spedding M, Langmead CJ. International union of basic and clinical pharmacology. XC. Multisite pharmacology: Recommendations for the nomenclature of receptor allosterism and allosteric ligands. Pharmacological Reviews. 2014;**66**:918-947. DOI: 10.1124/pr.114.008862

[43] Christopoulos A. Advances in G protein-coupled receptor allostery: From function to structure. Molecular Pharmacology. 2014;**86**:463-478. DOI: 10.1124/mol.114.094342

[44] Lane JR, Donthamsetti P, Shonberg J, Draper-Joyce CJ, Dentry S, Michino M, Shi L, López L, Scammells PJ, Capuano B, Sexton PM, Javitch JA, Christopoulos A. A new mechanism of allostery in a G protein-coupled receptor dimer. Nature Chemical Biology. 2014;**10**:745-752. DOI: 10.1038/nchembio.1593

[45] Gomes I, Ayoub MA, Fujita W, Jaeger WC, Pfleger KDG, Devi LA. G protein-coupled receptor heteromers. Annual Review of Pharmacology and Toxicology. 2016;**56**:10.1-10.23. DOI: 10.1146/annurev-pharmtox-011613-135952

[46] Kamiya T, Saitoh O, Yoshioka K, Nakata H. Oligomerization of adenosine A₂ₐ and dopamine D₂ receptors in living cells. Biochemical and Biophysical Research Communications. 2003;**306**:544-549. DOI: 10.1016/S0006-291X(03)00991-4

[47] Ferré S, Casadó V, Devi LA, Filizola M, Jockers R, Lohse MJ, Milligan G, Pin J-P, Guitart X. G protein-coupled receptor oligomerization revisited: Functional and pharmacological perspectives. Pharmacological Reviews. 2014;**66**:413-434. DOI: 10.1124/pr.113.008052

[48] Acuner Ozbabacan SE, Engin HB, Gursoy A, Keskin O. Transient protein-protein inter-actions. Protein Engineering, Design & Selection. 2011;**24**:635-648. DOI: 10.1093/protein/gzr025

[49] Perkins JR, Diboun I, Dessailly BH, Lees JG, Orengo C. Transient protein-protein interactions: Structural, functional, and network properties. Structure. 2010;**18**:1233-1243. DOI: 10.1016/j.str.2010.08.007

[50] Furlong TJ, Pierce KD, Selbie LA, Shine J. Molecular characterization of a human brain adenosine A$_2$ receptor. Brain Research. Molecular Brain Research. 1992;**15**:62-66. DOI: 10.1016/0169-328X(92)90152-2

[51] Venkatakrishnan AJ, Deupi X, Lebon G, Tate CG, Schertler GF, Babu MM. Molecular signatures of G-protein-coupled receptors. Nature. 2013;**494**:185-194. DOI: 10.1038/nature11896

[52] Impagliazzo A, Milder F, Kuipers H, Wagner MV, Zhu X, Hoffman RM, van Meersbergen R, Huizingh J, Wanningen P, Verspuij J, de Man M, Ding Z, Apetri A, Kükrer B, Sneekes-Vriese E, Tomkiewicz D, Laursen NS, Lee PS, Zakrzewska A, Dekking L, Tolboom J, Tettero L, van Meerten S, Yu W, Koudstaal W, Goudsmit J, Ward AB, Meijberg W, Wilson IA, Radošević K. A stable trimeric influenza hemagglutinin stem as a broadly protective immunogen. Science 2015;**349**:1301-1306. DOI: 10.1126/science.aac7263

[53] Borroto-Escuela DO, Marcellino D, Narvaez M, Flajolet M, Heintz N, Agnati L, Ciruela F, Fuxe K. A serine point mutation in the adenosine A$_{2A}$R C-terminal tail reduces receptor heteromerization and allosteric modulation of the dopamine D$_2$R. Biochemical and Biophysical Research Communications. 2010;**394**:222-227. DOI: 10.1016/j.bbrc.2010.02.168

[54] Berque-Bestel I, Lezoualc'h F, Jockers R. Bivalent ligands as specific pharmacological tools for G-protein-coupled receptor dimers. Current Drug Discovery Technologies. 2008;**5**:1-7. DOI: 10.2174/157016308786733591

[55] Joh NH, Wang T, Bhate MP, Acharya R, Wu Y, Grabe M, Hong M, Grigoryan G, DeGrado WF. De novo design of a transmembrane Zn^{2+}-transporting four-helix bundle. Science. 2014;**346**:1520-1524. DOI: 10.1126/science.1261172

[56] Song WJ, Tezcan FA. A designed supramolecular protein assembly with *in vivo* enzymatic activity. Science. 2014;**346**:1525-1528. DOI: 10.1126/science.1259680

[57] Gonen S, DiMaio F, Gonen T, Baker D. Design of ordered two-dimensional arrays mediated by noncovalent protein-protein interfaces. Science. 2015;**348**:1365-1368. DOI: 10.1126/science.aaa9897

[58] Huang J, Kang BH, Pancera M, Lee JH, Tong T, Feng Y, Imamichi H, Georgiev IS, Chuang G-Y, Druz A, Doria-Rose NA, Laub L, Sliepen K, van Gils MJ, de la Peña AT, Derking R, Klasse P-J, Migueles SA, Bailer RT, Alam M, Pugach P, Haynes BF, Wyatt RT, Sanders RW, Binley JM, Ward AB, Mascola JR, Kwong PD, Connors M. Broad and potent HIV-1 neutralization by a human antibody that binds the gp41-gp120 interface. Nature. 2014;**515**:138-142. DOI: 10.1038/nature13601

[59] Fibriansah G, Ibarra KD, Ng T-S, Smith SA, Tan JL, Lim X-N, Ooi JS, Kostyuchenko VA, Wang J, de Silva AM, Harris E, Crowe JE Jr, Lok S-M. Cryo-EM structure of an antibody that neutralizes dengue virus type 2 by locking E protein dimers. Science. 2015;**349**:88-91. DOI: 10.1126/science.aaa8651

[60] Gregorio GG, Masureel M, Hilger D, Terry DS, Juette M, Zhao H, Zhou Z, Perez-Aguilar JM, Hauge M, Mathiasen S, Javitch JA, Weinstein H, Kobilka BK, Blanchard SC. Single-molecule analysis of ligand efficacy in β_2AR-G-protein activation. Nature. 2017;**547**:68-73. DOI: 10.1038/nature22354

[61] Lewicki DN, Gallagher TM. Quaternary structure of coronavirus spikes in complex with carcinoembryonic antigen-related cell adhesion molecule cellular receptors. The Journal of Biological Chemistry. 2002;**277**:19727-19734. DOI: 10.1074/jbc.M201837200

[62] Scallon B, Cai A, Radewonuk J, Naso M. Addition of an extra immunoglobulin domain to two anti-rodent TNF monoclonal antibodies substantially increased their potency. Molecular Immunology. 2004;**41**:73-80. DOI: 10.1016/j.molimm.2004.01.006

[63] Gould HJ, Sutton BJ. IgE in allergy and asthma today. Nature Reviews Immunology. 2008;**8**:205-217. DOI: 10.1038/nri2273

[64] Wan T, Beavil RL, Fabiane SM, Beavil AJ, Sohi MK, Keown M, Young RJ, Henry AJ, Owens RJ, Gould HJ, Sutton BJ. The crystal structure of IgE-Fc reveals an asymmetrically bent conformation. Nature Immunology. 2002;**3**:681-686. DOI: 10.1038/ni811

[65] Jaakola V-P, Griffith MT, Hanson MA, Cherezov V, Chien EYT, Lane JR, Ijzerman AP, Stevens RC. The 2.6 angstrom crystal structure of a human A_{2A} adenosine receptor bound to an antagonist. Science. 2008;**322**:1211-1217. DOI: 10.1126/science.1164772

[66] Isberg V, de Graaf C, Bortolato A, Cherezov V, Katritch V, Marshall FH, Mordalski S, Pin J-P, Stevens RC, Vriend G, Gloriam DE. Generic GPCR residue numbers—Aligning topology maps while minding the gaps. Trends in Pharmacological Sciences. 2015;**36**:22-31. DOI: 10.1016/j.tips.2014.11.001

[67] McDermott G, Prince SM, Freer AA, Hawthornthwaite-Lawless AM, Papiz MZ, Cogdell RJ, Isaacs NW. Crystal structure of an integral membrane light-harvesting complex from photosynthetic bacteria. Nature. 1995;**374**:517-521. DOI: 10.1038/374517a0

[68] Nooren IM, Thornton JM. Diversity of protein-protein interactions. The EMBO Journal. 2003;**22**:3486-3492. DOI: 10.1093/emboj/cdg359

[69] Monod J, Wyman J, Changeux JP. On the nature of allosteric transitions: A plausible model. Journal of Molecular Biology. 1965;**12**:88-118. DOI: 10.1016/S0022-2836(65)80285-6

[70] Bouvier M. Oligomerization of G-protein-coupled transmitter receptors. Nature Reviews Neuroscience. 2001;**2**:274-286. DOI: 10.1038/35067575

[71] Dwyer ND, Troemel ER, Sengupta P, Bargmann CI. Odorant receptor localization to olfactory cilia is mediated by ODR-4, a novel membrane-associated protein. Cell. 1998;**93**:455-466. DOI: 10.1016/S0092-8674(00)81173-3

[72] Ahnert SE, Marsh JA, Hernández H, Robinson CV, Teichmann SA. Principles of assembly reveal a periodic table of protein complexes. Science. 2015;**350**:1331 (aaa2245-1-aaa2245-10). DOI: 10.1126/science.aaa2245

[73] Che Y, Fu A, Hou X, McDonald K, Buchanan BB, Huang W, Luan S. C-terminal processing of reaction center protein D1 is essential for the function and assembly of photosystem II in *Arabidopsis*. Proceedings of the National Academy of Sciences of the United States of America. 2013;**110**:16247-16252. DOI: 10.1073/pnas.1313894110

[74] Guskov A, Kern J, Gabdulkhakov A, Broser M, Zouni A, Saenger W. Cyanobacterial photosystem II at 2.9-Å resolution and the role of quinones, lipids, channels and chloride. Nature Structural & Molecular Biology. 2009;**16**:334-342. DOI: 10.1038/nsmb.1559

[75] Ferreira KN, Iverson TM, Maghlaoui K, Barber J, Iwata S. Architecture of the photosynthetic oxygen-evolving center. Science. 2004;**303**:1831-1838. DOI: 10.1126/science.1093087

[76] Deisenhofer J, Epp O, Miki K, Huber R, Michel H. Structure of the protein subunits in the photosynthetic reaction cetre of *Rhodopseudomonas viridis* at 3Å resolution. Nature. 1985;**318**:618-624. DOI: 10.1038/318618a0

[77] Niwa S, Yu LJ, Takeda K, Hirano Y, Kawakami T, Wang-Otomo Z-Y, Miki K. Structure of the LH1-RC complex from *Thermochromatium tepidum* at 3.0Å. Nature. 2014;**508**:228-232. DOI: 10.1038/nature13197

[78] Cogdell RJ, Gall A, Köhler J. The architecture and function of the light-harvesting apparatus of purple bacteria: From single molecules to *in vivo* membranes. Quarterly Reviews of Biophysics. 2006;**39**:227-324. DOI: 10.1017/S0033583506004434

[79] Cogdell RJ, Isaacs NW, Freer AA, Howard TD, Gardiner AT, Prince SM, Papiz MZ. The structural basis of light-harvesting in purple bacteria. FEBS Letters. 2003;**555**:35-39. DOI: 10.1016/S0014-5793(03)01102-5

[80] Hou Q, Dutilh BE, Huynen MA, Heringa J, Feenstra KA. Sequence specificity between interacting and non-interacting homologs identifies interface residues–a homodimer and monomer use case. BMC Bioinformatics. 2015;**16**:325. DOI: 10.1186/s12859-015-0758-y

[81] Launay G, Ceres N, Martin J. Non-interacting proteins may resemble interacting proteins: Prevalence and implications. Scientific Reports. 2017;**7**:40419. DOI: 10.1038/srep40419

[82] Bhattacharya S, Lam AR, Li H, Balaraman G, Niesen MJM, Vaidehi N. Critical analysis of the successes and failures of homology models of G protein-coupled receptors. Proteins. 2013;**81**:729-739. DOI: 10.1002/prot.24195

[83] Sapay N, Estrada-Mondragon A, Moreau C, Vivaudou M, Crouzy S. Rebuilding a macromolecular membrane complex at the atomic scale: Case of the Kir6.2 potassium channel coupled to the muscarinic acetylcholine receptor M2. Proteins. 2014;**82**:1694-1707. DOI: 10.1002/prot.24521

[84] Leman JK, Ulmschneider MB, Gray JJ. Computational modeling of membrane proteins. Proteins. 2015;**83**:1-24. DOI: 10.1002/prot.24703

Chapter 5

Calcitonin-Related Polypeptide Alpha Gene Polymorphisms and Related Diseases

Nevra Alkanli, Arzu Ay and Suleyman Serdar Alkanli

Additional information is available at the end of the chapter

http://dx.doi.org/10.5772/intechopen.78320

Abstract

Calcitonin gene-related peptide (CGRP) is a neuropeptide containing 37 amino acids. CGRP is a potent vasodilator neuropeptide, which has protective mechanisms in physiological and pathological conditions. When released, CGRP is a peptide that is active in the cerebral circulation and interacts with the sympathetic nervous system. CGRP is very important in the treatment of cardiovascular diseases. In addition, CGRP, which is also associated with pain processes, has an important role in inflammation. Calcitonin-associated polypeptide alpha (CALCA), one of the isoforms of CGRP, functions through the wide CGRP receptors. Polymorphisms occurring in the CALCA gene are associated with diseases such as ischemic stroke, Parkinson's disease, ovarian cancer, bone mineral density, migraine, schizophrenia, manic depression, and essential hypertension. In this section, the information was given about CALCA gene, which is one of its isoforms of CGRP. In addition, CALCA gene polymorphisms and diseases associated with these gene polymorphisms have also been addressed.

Keywords: CGRP, CALCA, gene polymorphism, PCR, RFLP, diseases

1. Introduction

The CGRP family consists of calcitonin, adrenomedullin, adrenomedullin 2 (intermedin) and amylin. CGRP is a potent vasodilator neuropeptide and it acts through its receptors. CGRP has wide perivascular localization. It is known that sensory fibers to exhibit a wide innervation throughout the body, and CGRP is localized in these sensory fibers. CGRP is also localized in non-neuronal tissues apart from these sensory fibers. CGRP has important protective properties in physiological and pathological conditions. It plays an important role in the treatment of cardiovascular diseases since it has a role as a vascular protective factor. In addition, the sensory fibers contained in CGRP are associated with pain processes. For this reason, CGRP also plays an important role in migraine pathophysiology. Another condition that is effective of CGRP is inflammation. CGRP has CALCA and calcitonin-related polypeptide beta (CALCB) isoforms [1].

The human CALCA gene is localized on the chromosome 11 (11p15.2-p15.1). This gene codes the calcitonin and CGRP. CALCA gene contains 1 promoter and 6 exons. It is known that the polymorphisms that occur in this gene are related to various diseases. Several polymorphisms have been identified in the CALCA gene. It has been determined that these polymorphisms are related to cerebrovascular, neurodegenerative, psychiatric diseases and hypertension-connected conditions. The most common of the CALCA gene polymorphisms is the CALCA T692C gene polymorphism. Besides this polymorphism, there are also various polymorphisms of CALCA gene such as CALCA-1786T>C, CALCA-624 (T/C), and CALCA (I/D) [2, 3].

Several studies have been conducted to determine whether CALCA gene polymorphisms are genetic risk factors for various diseases. In some of these studies, a significant relationship was found between CALCA gene polymorphisms and disease development risks. However, studies were also found in which this relation is not determined [2].

As a result, in the studies carried out with different populations, different results were found. Findings acquired from these studies that carried out with different ethnic groups will be an important indicator that new treatments for these diseases can be developed.

2. Calcitonin gene-related peptide

CGRP is a neuropeptide produced in consequence of alternative RNA processing of the calcitonin gene and containing 37 amino acids. CGRP gene family is composed of calcitonin, adrenomedullin, adrenomedullin 2 (intermedin) and amylin. CGRP has two important isoforms as CALCA and CALCB (calcitonin-related polypeptide beta). These isoforms of CGRP have similar structures and biological activities. However, separate genes form them. CGRP is also composed of receptor activity modifying protein (RAMP) and calcitonin receptor-like receptor (CLR). RAMP is a protein that changes receptor activity. The CLR receptor is also another receptor bound to the RAMP receptor. CGRP is an extremely powerful vasodilator that has protective mechanisms important for physiological and pathological conditions. Firstly, CGRP released from sensory nerves includes pain pathways. It is a known fact that the sensory fibers contained in CGRP are also related to the pain processes. There are studies showing that CGRP antagonists play an important role in migraine and have the potential to treat migraine. The studies are found that demonstrate that CGRP antagonists alleviate migraine. Apart from this disease, it is also known to have effects on arthritis, skin disorders, diabetes and obesity. Therefore, CGRP is a very important peptide in mammalian biology. CGRP is localized in the sensory fibers, which exhibit an innervation throughout the body, mainly with extensive perivascular localization. These sensory fibers are known to have a dual role in sensory (nociceptive) and efferent (effector) function. The role of CGRP is unclear, but it is also localized in the lesser known neuronal tissues. Sympathetic leakage mediation of CGRP in the brain has been shown. However, when exogenous CGRP was administered to femtomolar doses to skin of human and animal species, CGRP appeared to have a vasodilatory effect. Vascular protective role of CGRP has been identified via studies in various animal models. Therefore, it has been suggested that CGRP may be an important peptide in the treatment of cardiovascular diseases. CGRP is a very important neuropeptide with various aspects. Firstly, when CGRP is released, it is found as active in the cerebral circulation. In addition to be a powerful vasodilator, it is known that there is a reciprocal interaction with the sympathetic system in the environment. In addition to these, very important role of CGRP is found in inflammation [1].

2.1. Structure of CGRP

The structure of CALCA resembles CALCB the other isoform of CGRP. CALCA isoform consists of four domains. The first domain consists of the first seven residues of the NH_2 terminus, and it forms a ring-like structure that is held together with a disulfide bridge. CGRP 8-37 is a CGRP antagonist that occurred from removal of this first domain. Domain 2, composing an alpha helix, occurs from 8 to 18 residues, and these residues constitute deletions that cause 50- to 100-fold decrease in affinity. Residues of 11 and 18 are found in the hydrophilic face of the alpha helix. These residues also play an important role in supporting high-affinity binding. Domain 3 is occurs from 19 to 27 residues, and it is formed from the beta or gamma twist. The fourth domain comprises COOH terminus, and it consists of residues inherit from 28 to 37. It is believed that Domain 4 is required to form a binding epitope, and this domain has two domain rotations. When species differences and structure-activity relationships for CGRP are investigated, various amino acids have been identified. In receptor binding and activation, quite important functions of these amino acids have been found [1] (**Figure 1** and **Table 1**).

2.2. Molecular genetics of CGRP

CALCA gene is localized on the chromosome 11 (11p15.2-p15.1), and it contains six exons. Exon I is an untranslated region. While the exon II encodes signal peptide, exon III encodes N-terminal propeptide. Calcitonin and CGRP are localized on exon IV and V. The untranslated exon VI is the part of the CALCA. All of six exons constitute the primary mRNA transcript and then calcitonin or CGRP mRNA is formed. In consequence of combining the first three exons with exons V and VI, mRNA containing the CGRP is produced. Exon V codes the CGRP. Exon VI encodes the 3' untranslated region of the CGRP mRNA besides the polyadenylation (polyA) signal. mRNA is translated to produce the pro-CGRP peptide which is cleaved in the conjugated dibasic amino acids and the CGRP is released as 37th amino acid. The structure of the CALCB gene on chromosome 11 is like that of the CALCA gene. However, the exon 4 has lack polyA and thus alternative binding is prevented. In the consequence of this gene transcription, only CGRP is produced [4] (**Figure 2**).

Figure 1. Amino acid residues of human CALCA and CALCB isoforms.

Amino acids	Three-letter abbreviation	One-letter abbreviation	DNA codons	Chemical structure
Alanine	Ala	A	GCT, GCC, GCA, GCG	
Glycine	Gly	G	GGT, GGC, GGA, GGG	
Isoleucine	Ile	I	ATT, ATC, ATA	
Leucine	Leu	L	CTT, CTC, CTA, CTG, TTA, TTG	
Proline	Pro	P	CCT, CCC, CCA, CCG	
Valine	Val	V	GTT, GTC, GTA, GTG	
Phenylalanine	Phe	F	TTT, TTC	
Tryptophan	Trp	W	TGG	
Tyrosine	Tyr	Y	TAT, TAC	
Aspartic acid	Asp	D	GAT, GAC	
Glutamic acid	Glu	E	GAA, GAG	
Arginine	Arg	R	CGT, CGC, CGA, CGG, AGA, AGG	
Histidine	His	H	CAT, CAC	
Lysine	Lys	K	AAA, AAG	
Serine	Ser	S	TCT, TCC, TCA, TCG, AGT, AGC	

Amino acids	Three-letter abbreviation	One-letter abbreviation	DNA codons	Chemical structure
Threonine	Thr	T	ACT, ACC, ACA, ACG	
Cysteine	Cys	C	TGT, TGC	
Methionine	Met	M	ATG	
Asparagine	Asn	N	AAT, AAC	
Glutamine	Gln	Q	CAA, CAG	

Table 1. The chemical structure and DNA codons of amino acid residues of human CALCA and CALCB isoforms.

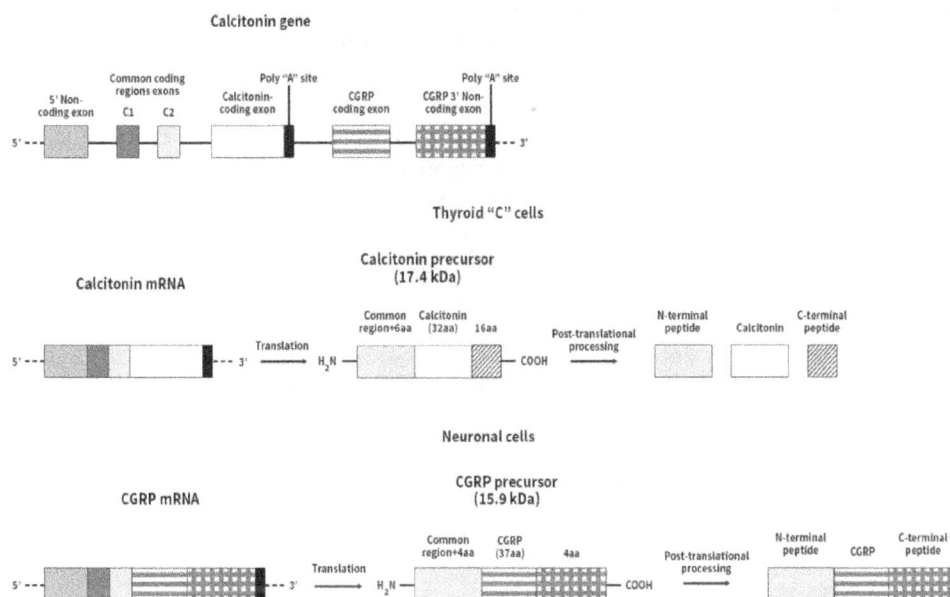

Figure 2. Calcitonin or CGRP production from CALC I gene.

2.3. Isoforms of CGRP

The CALCA and CALCB isoforms of CGRP are called as also CGRP I and CGRP II. These isoforms are synthesized from two different genes on chromosome 11 in different regions. The

CALCB is copied from the CALC II gene, the CALC I gene is to alternative binding to produce calcitonin or CGRP. The CALCA and CALCB sharing are analogous and 90% homologous, but they different from in terms of only three amino acids in humans. Therefore, the biological activities of these isoforms are similar. CALCA is the basic form that found in the central and peripheral nervous system. CALCB is the isoform that found in the enteric nervous system. Calcitonin is produced from the CALC I gene in consequence of expression in the mature protein of exon IV in the gene. Exon V and exon VI are converted to the 121st amino acid pro-hormone in consequence of the expression. CALCA is then cleaved to produce mature 37 amino acid peptides and mRNA is produced. The mechanism determining alternative binding for CALCA, which is predominantly expressed along the central, and peripheral nervous system, is still not fully understood [1].

2.4. Physiological functions of CGRP

2.4.1. CGRP in the cardiovascular system

The distribution of CGRP and its receptors in cardiac tissues such as the sinoatrial node, coronary arteries, atrial and ventricular muscle systems causes an increase in functions such as heart rate, contraction force, coronary heart flow and microvascular permeability. CGRP plays an important role in the regulation of vascular tone and angiogenesis. In consequence of CGRP infusion, perfusion pressure drops in isolated hearts, and vasodilator effect is observed in coronary vasculature. Also, CGRP shows a cardioprotective effect. Thus, capacitance blood vessels are directly affected and environmental vasodilation develops. CGRP receptors, which are found predominantly in the renal blood vessels, have various functions. These functions include increasing renal blood flow, increasing glomerular filtration rate, relieving glomerular afferent arterioles, increasing renin production, and stimulating arterial natriuretic peptide release [4].

2.4.2. CGRP in the central nervous system

CGRP plays a very important role in various functions such as motor, sensory and integrative systems in the central nervous system. Except this, CGRP is a peptide that modulates various senses. CGRP spreads in the central associated with autonomic functions. CGRP also plays an important role in regulating functions such as cardiovascular, respiratory and sleep functions. Apart from these functions, CGRP has a regulatory role. There is a regulatory effect of CGRP in the growth hormone release, hyperthermia, catalepsy, motor activity, and nociceptive responses. In addition, CGRP enhances excitatory actions by increasing the release of excitatory amino acid. CGRP that found in the efferent nerve fibers is found together with neurons containing acetylcholine. Thus, CGRP modulates the acetylcholine release. It also increases the synthesis of acetylcholine receptors and functions as a neurotrophic factor [4].

2.4.3. Other functions of CGRP

Other functions of CGRP are regulation of pituitary hormone secretion, release of pancreatic enzymes, control of gastric acid secretion, thermoregulation, reduction in food intake, insulin action, antagonism of insulin, growth factor-like functions. The CGRP effect is induced in the bones through the calcitonin receptors. In consequence of this induction, hypocalcemia, proliferation of osteoclasts, inhibition of both basal and stimulated absorption of the bone occurs. It is also known that CGRP is also distributed in bone tissues. In cases such as pregnancy, menstruation, or oral contraception, plasma CGRP levels increase. Spontaneous contractions occurring in uterus and fallopian tubes are also inhibited by CGRP effect. CGRP, which increases microvascular permeability, is also effective in the formation of inflammatory hyperemia, neutrophil accumulation, and localized edema. CGRP, which has the function of enhancing the migration of endothelial cells, plays an important role in the situations such as ischemia, inflammation, and wound healing [4].

3. Calcitonin-associated polypeptide alpha

3.1. Structure of CALCA gene

The human CALCA gene that encodes calcitonin and CGRP is localized on chromosome 11 (11p15.2-p15.1). CALCA gene that consisted of 1 promoter and 6 exons, performs its function through CGRP receptors [2] (**Figure 3**).

3.2. CALCA gene polymorphisms

Many gene polymorphisms are found that occurring in the CALCA gene. In some studies, it has been shown the polymorphisms in the CALCA gene to be associated with cerebrovascular diseases such as ischemic stroke. However, there are studies that show that CALCA gene polymorphisms are not a genetic risk factor for the development of ischemic stroke. Apart from ischemic stroke, it is known that CALCA gene polymorphisms may also be genetic risk factors for various diseases such as Parkinson's disease, ovarian cancer, bone mineral density, migraine, schizophrenia, and essential hypertension [5–8].

The most common polymorphism in the CALCA gene is the CALCA T692C gene polymorphism. Apart from this polymorphism, in the CALCA gene, -1786T>C, -624(T/C), 4218(T/C), -1784 (T/C), -1750 (C/G), -1218 (C/T), -1036 (G/A), rs7948017 (A/C), rs5241 (C/A), rs 2956 (A/T), -855 (G/A), -590 (C/G), CALCA (I/D), 2 bp microdeletion and CALCA A4218T>C gene polymorphisms are determined. Polymerase chain reaction (PCR) and restriction fragment length polymorphism (RFLP) methods are used to determine the CALCA gene polymorphisms genotype distributions [5–8].

Chromosome 11

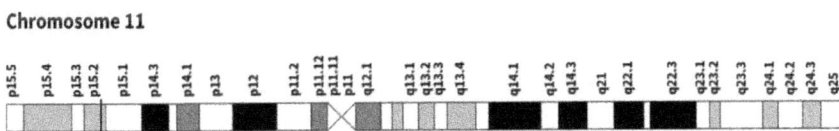

Figure 3. The structure of CALCA gene.

The CALCA T692C gene polymorphism is a single nucleotide polymorphism, and it is charac-terized by a T/C base transition in position 692 of the CALCA gene. In CALCA T692C gene polymorphism, three genotypes are observed as 692TT homozygote, 692CT heterozygote and 692CC homozygote. The forward primer for the CALCA T692C gene polymorphism is 5'-CGC ATC TGT ACC TTG CAA CT-3', and the reverse primer is 5'-TCA AAT TCC CGC TCA CTT TA-3'. The PCR conditions for the CALCA T692C gene polymorphism are 5 min for denatur-ation at 94°C, followed by 38 cycles of denaturation for 50 s at 94°C, annealing for 50 s at 57°C and extension for 1 min at 72°C, followed by 10 min of termination at 72°C. CALCA T692C gene polymorphism is determined using the restriction enzyme PshAI and product length is observed: 636 bp for the TT genotype; 636, 235, 401 bp for the CT genotype; 235, 401 bp for the CC genotype [2].

The CALCA-1786T>C gene polymorphism is another single-nucleotide gene polymorphism that belongs to the CALCA gene. This polymorphism is in the promoter region and arises in consequence of a T/C base exchange in position-1786. TT, CT and CC genotypes are observed in the CALCA-1786T>C gene polymorphism. For the CALCA-1786T>C gene polymorphism, the forward and reverse primers are 5'-CGC TGG GCT GTT TCT CAC AAT AT-3' and 5'-GTT AGA CAG GAG TTC AAT TAC AGT TGG C-3'. The PCR conditions for the CALCA-1786T>C gene polymorphism are 5 min for denaturation at 94°C, followed by 38 cycles of denaturation for 45 s at 94°C, annealing for 40 s at 62°C and extension for 45 s at 72°C, followed by 10 min of extension at 72°C. The genotype distributions of CALCA-1786T>C gene polymorphism are determined by using BsmAI restriction enzyme and product length is observed: 144 bp for the TT genotype; 144, 115, 29 bp for the CT genotype; 115, 29 bp for the CC genotype [2].

The CALCA 624 (T/C) gene polymorphism occurs because of a T/C base transition in position -624 of the CALCA gene promoter region. TT, CT and CC genotypes are observed in this polymorphism. The forward primer for the CALCA-624 (T/C) gene polymorphism is 5'-GCT GTT TCT CAC AAT ATT CC-3' and the reverse primer is 5'-CAA TTC CTG GTT GTG TGA TC-3'. For CALCA-624 (T/C) gene polymorphism, the PCR conditions are 10 min for denatur-ation at 94°C, followed by 35 cycles denaturation for 45 s at 95°C, annealing for 45 s at 60°C, and extension for 45 s at 72°C, followed by 7 min of termination at 72°C. The genotype distributions of CALCA-624 (T/C) gene polymorphisms are determined by using BsmAI restriction enzyme and product length is observed: 109 bp for the TT genotype; 109, 86, 23 bp for the CT genotype; 86, 23 bp for the CC genotype [2].

The forward primer for the CGRP 4218 (T/C) gene polymorphism is 5'-GGA AGA AGC AAA GAC CAG GA-3' and the reverse primer is 5'-CTG CAA GAA CAA TTC CCA CA-3'. The genotype distributions of CGRP 4218 (T/C) gene polymorphisms are determined by using AluI restriction enzyme and product length is observed: 202, 169, 106 bp for the TT genotype; 371, 202, 169, 106 bp for the CT genotype; 371, 106 bp for the CC genotype [9].

TT, CT and CC genotypes are observed in CALCA A4218T>C gene polymorphism. Primer sequences are used to determine this gene polymorphism; forward primer: 5'-AGC CTG CAC TGA GTT TGC TTC CC-3' and reverse primer: 5'-ATC CAC CTT CCT GTG TAT TGC TG CG-3. For CALCA A4218T>C gene polymorphism, the PCR conditions are 10 min for denaturation at 95°C, followed by 35 cycles denaturation for 45 s at 95°C, annealing for 45 s at 60°C, and extension for 45 s at 72°C, followed by 7 min of termination at 72°C. The genotype distribu-tions of CGRP 4218 (T/C) gene polymorphisms are determined by using AluI restriction enzyme and product length is observed: 140, 96 bp for the TT genotype; 236, 140, 96 bp for the CT genotype; 236, 96 bp for the CC genotype [10].

II, ID and DD genotypes are observed in the CALCA (I/D) gene polymorphism. Primer sequences are used to determine this gene polymorphism; forward primer: 5′-TTG GGG AGA AGG GTA GGA CT-3′ and reverse primer: 5′-GAA CTT TTG GAA GCC CAT GA-3. For CALCA (I/D) gene polymorphism, the PCR conditions are 10 min for denaturation at 95°C, followed by 30 cycles denaturation for 45 s at 95°C, annealing for 45 s at 60°C, and extension for 45 s at 72°C, followed by 4 min of termination at 72°C. Product lengths in the CALCA (I/D) gene polymorphism are 303 bp (wildtype-I) and 287 bp (deletion-D) [6].

CC, CG and GG genotypes are observed in the CALCA-1750 (C/G) gene polymorphism. Primer sequences are used to determine CALCA-1750 (C/G) gene polymorphism; forward primer: 5′-TAG CTG GTA TTT CCC ACA GAG-3′ and reverse primer: 5′-CCC ATT TCA AAG ATG AGT ACC CTG-3. The genotype distributions of CALCA-1750 (C/G) gene poly-morphisms are determined by using Bsu36I restriction enzyme and product length is observed: 167 bp for the GG genotype; 167, 142, 25 bp for the CG genotype; 142, 25 bp for the CC genotype [3].

Primer sequences are used to determine CALCA 2 bp microdeletion gene polymorphism; forward primer: 5′-CCC AGA AGA GGA GGA CAG CTC TGG GT-3′ and reverse primer: 5′-AGA GCT GGA GGA GCG ATC CTA GAG GGA-3. For CALCA (I/D) gene polymorphism, the PCR conditions are 3 min for denaturation at 96°C, followed by 60 cycles denaturation for 25 s at 98°C, annealing for 30 s at 63°C, and extension for 30 s at 72°C, followed by 10 min of termination at 72°C. Product lengths in the CALCA 2 bp microdeletion gene polymorphism are 184 and 182 bp [11].

TT, CT, and CC genotypes are observed in the CALCA-1218 (C/T) gene polymorphism. Primer sequences are used to determine this gene polymorphism; forward primer: 5′-CAG GTT CTG GAA GCA TGA GGG TGA CGC′ and reverse primer: 5′-CGA CTG CTC TTA TTC CCG CCG CTG T-3. For CALCA-1218 (C/T) gene polymorphism, the PCR conditions are 3 min for denaturation at 96°C, followed by 60 cycles denaturation for 25 s at 98°C, annealing for 30 s at 63°C, and extension for 30 s at 72°C, followed by 10 min of termination at 72°C [11].

GG, GA and AA genotypes are observed in the CALCA-855 (G/A) gene polymorphism. The wild-type sequence used to determine this gene polymorphism is: 5′-GGC TTC CGC ATC TGTA-3′ and mutation sequence: 5′-GGC TTC CAC ATC TGTA-3′. For CALCA-855 (G/A) gene polymorphism, the PCR conditions are 35 cycles denaturation for 40 s at 94°C, annealing for 45 s at 56°C, and extension for 1 min at 72°C. The genotype distributions of CALCA-855 (G/A) gene polymorphisms are determined by using AciI restriction enzyme [10].

CC, CG and GG genotypes are observed in the CALCA-590 (C/G) gene polymorphism. The wild type and mutation sequences are used to determine the polymorphism of this gene respectively as 5′-ACA CTG AGC CTC TGT-3′ and 5′-ACA CTC AG**G** CTC TGT-3′. For CALCA-590 (C/G) gene polymorphism, the PCR conditions are 35 cycles denaturation for

40 s at 94°C, annealing for 45 s at 56°C, and extension for 1 min at 72°C. The genotype distributions of CALCA-590 (C/G) gene polymorphisms are determined by using PshAI restriction enzyme [10].

3.3. Migraine and CALCA gene polymorphisms

Migraine, a common disease, is characterized by unilateral throbbing headache with autonomic symptoms such as nausea, vomiting, and photophobia. Although the pathogenesis of migraine is still unclear, it is known that genetic and environmental factors play a role in the pathophysiology of this disease [12].

At the onset of migraine attack, trigeminovascular system is activated. Vasodilatation occurs in the cranial blood vessels in consequence of the release of substance P, neurokinin A and CGRP at sensory nerve endings. As a result, neurogenic inflammation occurs in these veins. Pain signals are induced, and they transmitted to the thalamus. These signals are perceived as headache by the cerebral cortex [12].

Increased levels of CGRP obtained from jugular vein are related to the development of migraine attacks. These CGRP levels return to normal following headache interruption. Intravenous infusion of CGRP is effective in the formation of a like migraine headache. When properly administered, CGRP antagonists can prevent migraine attacks. Nitric oxide is another substance that plays an important role in the pathogenesis of migraine, and nitric oxide effect is seen in consequence of CGRP release in the trigeminal nerve terminals [12].

CGRP is very important in migraine pathophysiology. CGRP is a peptide, which is responsible for neurological inflammation and vasodilatation in head trauma. It plays an important role in the regulation of vascular tone and angiogenesis by causing vasodilatation in blood vessels. It is known to be a neurotrophic factor modulating pain sensation in the nervous system. CGRP, an important peptide, must be synthesized correctly for biological activities to be regular. The molecular structure, function and reaction can change in consequence of the polymorphisms occurring in the CALCA gene [12].

Many studies have been conducted to investigate the relationship between migraine development risk and CALCA gene polymorphisms. In a study conducted in the Thracian population, CALCA T692C gene polymorphism genotype and allele, distributions in female migraine patients were not determined different from healthy controls. This polymorphism was found not to be associated with severity and frequency of migraine attacks. The significant difference was not found in terms of CALCA T692C gene polymorphism in comparison carried out between migraine types with and without aura [12].

In a study conducted by Menon et al., in the Australian population, the significant difference was not found in terms of CALCA (I/D) gene polymorphism between migraine patients (migraine with aura-migraine without aura) and controls. In a study performed by Lemos et al., the significant difference was not determined in terms of CALCA-1750 (C/G) gene polymorphism between migraine patients (with and without aura) and controls. In this study, it was also found that the coexistence of the CG genotype of CALCA-1750 (C/G) gene polymorphism the AT genotype of the brain natriuretic factor gene polymorphism of increased the risk of the resulting migraine [12].

In a study conducted in the Han-Chinese population, no significant relationship was found between CALCA rs 3781719 and rs 145837941 gene polymorphisms and the risk of developing migraine. However, in this study, CALCA rs 3781719 gene polymorphism was an important risk factor for the development of migraine with aura, but significant result was not found statistically. In a study conducted in the Japanese population by Masakazu et al., it was determined that CALCA rs 3781719 and rs 145837941 gene polymorphisms were not genetic risk factors for migraine complications due to excessive drug use in migraine patients. In another study conducted by Sutherland et al. in the Australian Caucasian population, there was no relationship between CALCA rs 3781719 and rs 145837941 gene polymorphisms and the risk of developing migraine. In the study carried out by Lemos et al., in a European population, CALCA rs 1553005 gene polymorphism was not found as a genetic risk factor for the development of migraine [12].

3.4. Essential hypertension and CALCA gene polymorphisms

Essential hypertension, which affects about 20–25% of the world's population, is a very important health problem. It is known that hypertension increases the risk of coronary heart disease, ischemic stroke, and congestive heart failure. Essential hypertension is a multifactorial disease, and it is quite complex. Environmental and genetic factors play a role in the development of essential hypertension. Through changes in CGRP synthesis and release, CGRP plays an important role in the onset, progression of essential hypertension and its maintenance of essential hypertension [7].

Several single nucleotide gene polymorphisms have been found as effective in the development of essential hypertension. It has been determined that some polymorphisms that occur in the genes are important in influencing the expressions of enzymes and proteins associated with essential hypertension such as angiotensinogen, endothelial nitric oxide synthase. CGRP is a neuropeptide that plays an important role in the pathophysiology of essential hypertension. CALCA and CALCB isoforms of CGRP are associated with increasing of blood pressure. Differences in CGRP plasma concentrations were not fully determined between healthy subjects and hypertensive patients. However, significantly lower plasma CGRP concentrations were found in hypertensive patients and preeclamptic pregnant women than normotensive controls. In some studies performed in patients with hypertension, a significant relationship was found between elevated plasma CGRP levels and systolic and diastolic blood pressures [7].

In consequence of the polymorphisms occurring in the CALCA gene, heart diseases and renal damage due to hypertension are also increasing. In some studies performed with experimental animals, it was observed that systolic blood pressure increases in consequence of CALCA gene polymorphisms. In a study conducted in Japan, a 2-bp microdeletion polymorphism has been shown in the intron 1 of the CALCA gene. This gene polymorphism is associated with the risk of developing essential hypertension. In another study conducted with the Chinese population, it was determined that CALCA T692C gene polymorphism is a genetic risk factor in the development of essential hypertension [7].

3.5. Ischemic stroke and CALCA gene polymorphisms

Cerebrovascular diseases occur in consequence of sudden emergence of local or global neurological symptoms. Ischemic stroke is one of these diseases, and it occurs in consequence of blocking of blood flow to any region of the brain. Ischemic stroke is classified into five subgroups. The emergence of large or small vessel diseases is a major cause of cerebral ischemia. Large artery disease arises due to atherothrombosis of the carotid, vertebral and proximal cerebral arteries. Lipohyalinosis of the vessel wall in the distal penetrant branches of the veins results in small artery disease. Another ischemic stroke subtype arising from endometrium diseased cardiac valves is cardioembolism. In consequence of these pathogenic mechanisms, a decline is observed in cerebral blood flow. In consequence of decrease the levels of oxygen and glucose required to feed of the brain reduce cell damage occurs. Another type of stroke that causing hypercoagulability is cryptogenic stroke. The other one is also an unclassified ischemic stroke subtype [2].

CGRP is an important member of the calcitonin peptide family, and it plays an important role in the dilation of the cerebral arteries in the human. It is a neuropeptide that especially associated with central and peripheral nervous system disorders. It is known that the polymorphisms occurring in the CALCA gene is associated with ischemic stroke. However, in some studies, it has been determined that CALCA gene polymorphisms were not to be genetic risk factors for the development of ischemic stroke. In a study conducted in the Thracian population, it was determined that CALCA T692C, -1786T>C and -624 (T/C) gene polymorphisms were not genetic risk factors for the development of ischemic stroke. In addition, the significant difference was not determined in terms of the CALCA gene polymorphisms in the patients' subtypes with ischemic stroke in the same study. A limited number of studies have found aimed to investigate the relationship between the risk of ischemic stroke development and CALCA gene polymorphisms [2].

3.6. Ovarian cancer and CALCA gene polymorphisms

Calcitonin, which is synthesized by parathyroid cells of the thyroid, is a peptide hormone that plays an important role in suppressing blood calcium levels. It is known that reducing extracellular calcium levels is associated with the regulation of calcitonin. The risk of ovarian cancer can be reduced by an antiproliferative mechanism. In the ovulation period, the ovarian surface epithelium is subjected to repeated injuries and healing. Repeated proliferative stimulation plays an important role in the malignant transformation of ovarian epithelial cells. Extracellular calcium is elevated via receptors that sensing the calcium. Unconverted ovarian surface epithelial cells multiply in response to this condition [13].

The differentiation and proliferation of small intestine and chest epithelial cells are also modulated by extracellular calcium. There is a very significant relationship between this differentiation, proliferation and ovarian cancer. Therefore, impaired calcium regulation is believed to be an important risk factor for the development of ovarian cancer. Calcitonin, which plays an important role in the prevention of bone resorption, causes a decrease in serum calcium and is known as an important regulator of calcium metabolism by this property. Calcium inhibits the production of proteins necessary for the regulation of hypocalcemia. Increased calcium reduces proliferation and thus regulates wound healing of the ovarian surface epithelium [13].

Hypercalcemia is also associated with ovarian epithelial tumors. Proteins that are effective in the regulation of hypocalcemia stimulate protein kinase C and phospholipase C signaling pathways. Thus, mitosis is triggered and cancer-related apoptosis reduces. It can also alter the cancer susceptibility by interacting with insulin-like growth factor (IGF) and binding protein, hormones and other growth factors [13].

In studies conducted in the Japanese population, it was found that the C allele of CALCA-624 (T/C) gene polymorphism associated significantly with ovarian cancer. In women with TT genotypes, ovarian cancer is much more common than in women with CC genotype. It has been determined that the C allele is a genetic risk factor in women who consume less calcium. In addition, because of reduced calcium intake, serum calcium levels decrease and the concentration of required calcitonin decreases [13].

3.7. Parkinson, schizophrenia, manic depression and CALCA gene polymorphisms

Parkinson's disease occurs because of damage of gray matter nuclei in the lower part of the brain. Because of this damage, degeneration occurs in cells that secrete dopamine [14]. Schizophrenia, a chronic brain disease, is also known as a developmental disorder of the brain. It is known that subcortical dopamine pathways in the brain are highly active in this disease. In consequence of this over activity, excessive dose of dopamine is released from the nerve endings. As a result, dopamine receptors are stimulated largely [15]. The dopaminergic hypothesis plays an important role in the pathophysiology of manic depression, a common neuropsychiatric disorder. There are homeostatic mechanisms that develop in response to hyperdopaminergic agents in the manic phase of the disease [16]. Because of these mechanisms, decrease occurs in dopaminergic function together with a hypodopaminergic situation and depression can developed [10].

Calcitonin plays an important role in bone calcium metabolism and is produced by C cells known as thyroid parafollicular cells. Polymorphisms that occur in calcitonin receptors are associated with changes in bone density in postmenopausal women. It is believed that the disorders that occur in calcitonin function are related to osteoporosis. Concentrations of procalcitonin are increased in bacterial infections and in cases such as bacterial sepsis. This increase is performed as independent of thyroid C cells. Catacalsine (PND-21) is another product of procalcitonin and is important in the development of medullary thyroid carcinoma [10].

CALCA, which produces a strong vasodilator effect in the brain and peripheral organs, is released from nerve fibers that bind blood vessels to the nervous system. CALCA, released from fibrils that come from the ventral tegmental area, amygdala, and ventral striatum plays an important role as moderator of dopaminergic transmission in these areas. The dopaminergic system consists of mesolimbic and mesostral axis. Both axes are influenced by CALCA. There was a relationship between cerebrospinal fluid and elevated CALCA levels in patients with major depression. Polymorphisms of CALCA gene have been associated with disorders such as Parkinson, Schizophrenia, and manic depression. It is believed that gene polymorphisms of CALCA 855 (G/A), CALCA-624 (T/C), CALCA-590 (C/G) and CALCA (I/D) may be associated with psychiatric or neurological diseases in dopaminergic transmission [10].

3.8. Aseptic loosened total hip arthroplasty and CALCA gene polymorphisms

One of the most important causes of long-term failure of arthroplasty is aseptic loosening of total joint replacement. There are various studies to understand the aseptic relaxation mechanism. Aseptic relaxation is known as a multifaceted process resulting from events such as foreign body reaction, cell-cell interaction, allergic reactions, hydrostatic pressure, body weight, and implantation. Protein-surrounding local osteolysis is initiated by an aseptic inflammatory response following the addition of abrasion particles by the macrophages. Fibroblasts are stimulated by macrophages in consequence of proliferation and differentiation of precursor osteoclasts to mature osteoclasts. This stimulation performs via cytokines. An important process in the regulation of homeostasis in tissues is apoptosis. Some important proteins regulate the most important process and the proliferation of normal tissues.

In a study conducted by Ahmed et al., CALCA immunoreacted nerve fibers were found in the arthroplasty interface membrane. CALCA, produced by central and peripheral nervous systems, is found in almost all tissues and plays an important role in bone remodeling. Large quantities of calcitonin produced by thyroid cells are responsible for calcium homeostasis, leading to the inhibition of osteoclasts. Because of changes in cell cycle, apoptosis, receptor expression, DNA replication, gene expressions and genes, the risk of aseptic loosening following total hip replacement can be increased. In a study conducted in Germany, it was aimed to determine the relationship between aseptic loosening risk and CALCA-1786T>C gene polymorphism following total hip replacement. In consequence of this study, it was determined that CALCA-1786T>C gene polymorphism was not a genetic risk factor for the risk of aseptic loosening [8].

3.9. Analgesic effect and CALCA gene polymorphisms

In most of cancer patients, severe pain occurs in the terminal period. Opioid drugs, known as the most effective drugs in the treatment of cancer pain, can lead to tolerance and hyperalgesia when used for a long time. Fentanyl, one of the opioid drugs, is used as an analgesic in the treatment of terminal cancer pain. Recently, it has been shown that there is a significant relationship between fentanyl and human genes. It is known that CGRP is an important transmitter in the transmission of pain signals. The analgesic mechanism in opioid medications reduces the release of neurotransmitters such as CGRP into the synaptic cleft of nerve fiber ends of the presynaptic membrane [9].

In a study conducted by Cepeda et al., it is found that respiratory depression arising from morphine was found to be more effective in Native American Indians than Caucasians. In another study conducted by Zhou et al. in Caucasians, sedation and respiratory depression were found to be more observed than in Asians. When comparing men and women, it was also determined that the reactions to these drugs were different. In a study by Sear et al., no association was found between plasma concentrations of opioid drugs and clinical effects. The analgesic effect of fentanyl is different. It is believed that this diversity to related to gene motion modes [9].

In some studies, genetic variants that can affect the efficacy of fentanyl have been identified. Because of the polymorphisms that occur in the genes, the analgesic effect and the pain perception mechanism in humans are affected. CGRP is an important neuropeptide in the nervous system. This neuropeptide plays an important role in peripheral nerves via nociceptive information transfer. It is also effective in the production of hyperalgesia in the spinal cord [9].

As a newly discovered polymorphism in exon III of the CGRP gene, CGRP 4218 (T/C) gene polymorphism is believed to be associated with diseases such as Parkinson's and major depression. Whether there is an association between the analgesic effect of fentanyl and the CGRP 4218 (T/C), gene polymorphism is unclear. In a study, CGRP 4218 (T/C) gene polymorphism was identified as a risk factor for the analgesic effect of fentanyl. This study has proven that there is no association between CGRP 4218 (T/C) gene polymorphism and postoperative adverse reactions resulting from fentanyl. In addition to these studies, studies on the relationship between CGRP 4218 (T/C) gene polymorphism and fentanyl pharmacokinetics should be performed [9].

3.10. Psoriasis vulgaris and CALCA gene polymorphisms

The pathogenesis of psoriasis, a chronic disease characterized by reddish, scaly skin, is quite complicated. Intravascular molecules such as intracellular adhesion molecule (ICAM), TNF-alpha, reactive oxygen species (ROS) play an important role in the pathogenesis of this disease [17].

CGRP, widely distributed in central and peripheral nervous systems, is a peptide intermediating to pain. This peptide is also known as a growth factor in cells such as Schwan cells and endothelial cells. Proinflammatory features of CGRP are found in many diseases. In several earlier studies, a marked cutaneous proliferation has been identified in psoriasis. It has also been found that plaques of psoriasis include CGRP. Because of this situation, it has been shown that psoriatic progression is observed. In consequence of various studies, target genes have been identified which can contribute to the development of psoriasis. One of the isoforms of CGRP, which plays an important role in immune regulation and inflammation, is the CALCA gene. It is believed that polymorphisms in this gene may be associated with the development of psoriasis [17].

In a study conducted with the Chinese population, the effect of CGRP mRNA expression and plasma CGRP levels on the development of psoriasis was investigated. It was also aimed to determine the relationship between the T692C gene polymorphism in the CALCA gene and the risk of development of psoriasis. In this study, morbidity of psoriasis was found to be increased CGRP expression and release. Furthermore, in the CALCA T692C gene polymorphism, TT genotype has been identified as a genetic risk factor in the development of psoriasis in people with alcohol habits. Ethanol that can activate primer sensory neurons can cause neuropeptide release in the skin. In consequence, activation capsaicin receptor of the alcohol results in the release of CGRP from the sensory nerves [17].

It is known that CGRP, a potent vasodilator, plays an important role in skin homeostasis. In consequence of intradermal injection of CGRP, increased blood flow is observed, and microvascular dilation is induced. Consequently, local erythema occurs. There are not many studies aiming to investigate the relationship between CALCA gene polymorphisms and psoriasis [17].

4. Conclusion

Polymorphisms occurring in the CALCA gene are known to be associated with various diseases. Several studies have been carried out to investigate the relationship between CALCA gene polymorphisms and some diseases. In some of these studies, it has been determined that CALCA gene polymorphisms are genetic risk factors in the development of these diseases. However, there are also studies showing that there is no significant relationship between CALCA gene polymorphisms and these diseases. Different results can be obtained in studies; this situation may arise from different selection criteria for patients and control groups. In addition, the studies performed with a small number of cases are another reason for the different results. Different results were found in gene polymorphism studies carried out with different racial and ethnic populations. Findings obtained in consequence of carrying out these studies with more cases and with different populations will be an important indicator for the treatment of diseases.

Acknowledgements

Nevra Alkanli, Arzu Ay and Suleyman Serdar Alkanli carried out this chapter in department of Biophysics of T.C. Halic University Medical Faculty, Trakya University Medical Faculty and Istanbul University Medical Faculty.

Conflict of interest

We declare that there is no conflict of interest with any financial organization regarding the material discussed in the chapter.

Author details

Nevra Alkanli[1]*, Arzu Ay[2] and Suleyman Serdar Alkanli[3]

*Address all correspondence to: nevraalkanli@halic.edu.tr

1 Department of Biophysics, Faculty of Medicine, T.C. Halic University, Istanbul, Turkey

2 Department of Biophysics, Faculty of Medicine, Trakya University, Edirne, Turkey

3 Department of Biophysics, Faculty of Medicine, Istanbul University, Istanbul, Turkey

References

[1] Russell FA, King R, Smillie S-J. Calcitonin gene-related peptide: Physiology and patho-physiology. Physiological Reviews. 2014;**94**:1099-1142. DOI: 10.1152/physrev.00034.2013

[2] Alkanli N, Sipahi T, Ay A. Calcitonin related polypeptide alpha gene polymorphisms according to plasma total homocysteine levels in ischemic stroke patients of Trakya Region. Biotechnology & Biotechnological Equipment. 2017;**31**(6):1184-1191. DOI: 10.10 80/13102818.2017.1372218

[3] Ishii M, Katoh H, Kurihara T. Lack of association between CGRP-related gene polymor-phisms and medication overuse headache in migraine patients. Integrative Molecular Medicine. 2015;**2**(1):92-94. DOI: 10.15761/IMM.1000117

[4] Udayasankar A. Calcitonin gene-related peptide and migraine: implications for therapy [thesis]. Rotterdam: Erasmus University; 2004

[5] Sutherland HG, Buteri J, Menon S. Association study of the calcitonin gene-related polypeptide-alpha (CALCA) and the receptor activity modifying 1 (RAMP1) genes with migraine. Gene. 2013;**515**:187-192

[6] Menona S, Buteria J, Roy B. Association study of calcitonin gene-related polypeptide-alpha (CALCA) gene polymorphism with migraine. Brain Research. 2011;**1378**:119-124

[7] Xin-lin L, Tian-lun Y, Xiao-ping C, et al. Association of CALCA genetic polymorphism with essential hypertension. Chinese Medical Journal-Peking. 2008;**121**(15):1407-1410

[8] Wedemeyer C, Kauther MD, Hanenkamp S. BCL2-938C>A and CALCA-1786T>C poly-morphisms in aseptic loosened total hip arthroplasty. European Journal of Medical Research. 2009;**14**:250-255

[9] Yi Y, Zhao M, Xu F. CGRP 4218T/C polymorphism correlated with postoperative analge-sic effect of fentanyl. International Journal of Clinical and Experimental Pathology. 2015; **8**(5):5761-5767

[10] Buervenich S, Xiang F, Sydow O. Identification of four novel polymorphisms in the calcitonin/a-CGRP (CALCA) gene and an investigation of their possible associations with Parkinson disease schizophrenia, and manic depression. Human Mutation. 2001;**17**(5): 435-436. Mutation in Brief #416. Online WILEY–LISS, INC

[11] Morita A, Nakayama T, Soma M. Association between the calcitonin-related peptide α (CALCA) gene and essential hypertension in Japanese subjects. American Journal of Hypertension. 2007;**20**:527-532. DOI: 10.1016/j.amjhyper.2006.06.008

[12] Guldiken B, Sipahi T, Tekinarslan R, et al. Calcitonin gene related peptide gene polymorphism in migraine patients. The Canadian Journal of Neurological Sciences. 2013;**40**: 722-725. DOI: 10.1017/S0317167100014980

[13] Goodman MT, Ferrell R, McDuffie K. Calcitonin gene polymorphism CALCA-624 (T/C) and ovarian cancer. Environmental and Molecular Mutagenesis. 2005;**46**:53-58. DOI: 10.1002/em.20134

[14] Akbayir E, Şen M, Ay U. Parkinson hastalığının etyopatogenezi (Etiopathogenesis of Parkinson's disease). Deneysel Tıp Araştırma Enstitüsü Dergisi. 2017:**1-23**;1-23

[15] Acar C, Kartalci S. The role of catechol-O-methyltransferase (COMT) gene in the etiopathogenesis of schizophrenia. Current Approaches in Psychiatry. 2014;**6**(3):217-226. DOI: 10.5455/cap.20131025120419

[16] Ashok AH, Marques TR, Jauhar S. The dopamine hypothesis of bipolar affective disorder: The state of the art and implications for treatment. Molecular Psychiatry. 2017:**1-23**;1-23. DOI: 10.1038/mp.2017.16

[17] Guo R, Li FF, Chen ML. The role of CGRP and CALCA T-692C single-nucleotide polymorphism in psoriasis vulgaris. Die Pharmazie. 2015;**70**:88-93. DOI: 10.1691/ph.2015.4722

www.ingramcontent.com/pod-product-compliance
Lightning Source LLC
Chambersburg PA
CBHW081240190326
41458CB00016B/5851